Statistical and Machine Learning Models for Remote Sensing Data Mining - Recent Advancements

Statistical and Machine Learning Models for Remote Sensing Data Mining - Recent Advancements

Editors

Monidipa Das
Soumya K. Ghosh
V. M. Chowdary
Pabitra Mitra
Santosh Rijal

MDPI • Basel • Beijing • Wuhan • Barcelona • Belgrade • Manchester • Tokyo • Cluj • Tianjin

Editors

Monidipa Das
Machine Intelligence Unit
Indian Statistical Institute
Kolkata
India

Soumya K. Ghosh
Department of Computer
Science and Engineering
Indian Institute of Technology
Kharagpur
Kharagpur
India

V. M. Chowdary
Department of Agriculture,
Cooperation Farmers Welfare
Mahalanobis National Crop
Forecast Centre
New Delhi
India

Pabitra Mitra
Department of Computer
Science and Engineering
Indian Institute of Technology
Kharagpur
Kharagpur
India

Santosh Rijal
Department of Geography
Virginia Tech
Blacksburg
United States

Editorial Office
MDPI
St. Alban-Anlage 66
4052 Basel, Switzerland

This is a reprint of articles from the Special Issue published online in the open access journal *Remote Sensing* (ISSN 2072-4292) (available at: www.mdpi.com/journal/remotesensing/special_issues/rs_data_mining).

For citation purposes, cite each article independently as indicated on the article page online and as indicated below:

LastName, A.A.; LastName, B.B.; LastName, C.C. Article Title. *Journal Name* **Year**, *Volume Number*, Page Range.

ISBN 978-3-0365-4592-9 (Hbk)
ISBN 978-3-0365-4591-2 (PDF)

Contents

About the Editors

Monidipa Das

Monidipa Das is currently a DST-INSPIRE Faculty at the Machine Intelligence Unit (MIU), in the Indian Statistical Institute (ISI) Kolkata, India. Previously she was a postdoctoral research fellow in the School of Computer Science and Engineering (SCSE), Nanyang Technological University (NTU), Singapore. She received her Ph.D. degree in computer science and engineering from the Indian Institute of Technology (IIT) Kharagpur, in 2018, and her M.E. degree in computer science and engineering from the Indian Institute of Engineering Science and Technology (IIEST), Shibpur, in 2013. Her research interests include spatial informatics, spatio-temporal data mining, soft computing, and machine learning. Dr. Das has research publications in a number of revered international journals and international conferences. Dr. Das is member of the ACM, the IEEE Computational Intelligence Society, and the IEEE Geoscience and Remote Sensing Society.

Soumya K. Ghosh

Soumya K. Ghosh received his Ph.D. degree in Computer Science and Engineering from the Indian Institute of Technology (IIT) Kharagpur, India. Presently, he is a Professor with the Department of Computer Science and Engineering, IIT Kharagpur. Before joining IIT Kharagpur, he worked for the Indian Space Research Organization in the area of satellite remote sensing and geographic information systems. He has been awarded the National Geospatial Chair Professorship by the Department of Science and Technology, Government of India, in 2017. He has more than 300 research papers in reputed journals and conference proceedings. His research interests include spatial data science, spatial web services, and cloud computing. Dr. Ghosh is a senior member of IEEE.

V. M. Chowdary

V. M. Chowdary is currently the Director of the Mahalanobis National Crop Forecast Centre (MNCFC), Department of Agriculture, Cooperation & Farmers Welfare. Previously, he was the Deputy General Manager, Regional Remote Sensing Center –North (Delhi), NRSC, ISRO. Earlier, he also worked as a scientist and head (applications) at the Regional Remote Sensing Centre-East, Kolkata, National Remote Sensing Centre (NRSC), ISRO, India. His areas of interest include: the application of geospatial technologies, multi-criteria analysis and soft computing tools for agricultural water management, integrated watershed management, hydrological modeling, and land use/cover changes. He has published widely in peer-reviewed journals. He is the recipient of the Eminent scientist award from NESA and Team excellence award from ISRO. He has also been selected as the 'Fellow of A. P. Akademi of Sciences-FAPAS'for the year 2019.

Pabitra Mitra

Pabitra Mitra received the B.Tech. degree in electrical engineering from IIT Kharagpur, Kharagpur, India, in 1996, and the Ph.D. degree from the Department of Computer Science and Engineering, Indian Statistical Institute, Kolkata, India, in 2005. He is currently working as a Professor with the Department of Computer Science and Engineering, IIT Kharagpur. His research focuses on machine learning, pattern recognition, data mining, information retrieval, and computational biology. He is a recipient of prestigious Indian National Academy of Engineering Young Engineer Award and Royal Society UK Indian Science Network Award. He is a member of the Indian Unit for Pattern Recognition and Artificial Intelligence. He is a senior member of IEEE.

Santosh Rijal

Dr. Santosh Ris an assistant professor in the department of geography at Virginia Polytechnic Institute and State University (Virginia Tech). He completed his Masters degree from University of North Dakota and his Doctoral degree from Southern Illinois University Carbondale. Dr. R's research is focused on the application of geospatial technology in land condition disturbance, land use/land cover change, and natural resource management. He has been involved in research related to assessing and monitoring military installations land condition in the United States. He currently teaches geospatial courses such as Principles of GIS, Modeling with GIS, Introduction to Remote Sensing, etc., in the geography department at Virginia Tech.

Editorial

Statistical and Machine Learning Models for Remote Sensing Data Mining—Recent Advancements

Monidipa Das [1,*], Soumya K. Ghosh [2], Vemuri M. Chowdary [3], Pabitra Mitra [2] and Santosh Rijal [4]

1 Faculty in the Machine Intelligence, Indian Statistical Institute, Kolkata 700108, India
2 Department of Computer Science and Engineering, Indian Institute of Technology Kharagpur, Kharagpur 721302, India; skg@cse.iitkgp.ac.in (S.K.G.); pabitra@gmail.com (P.M.)
3 Mahalanobis National Crop Forecast Centre, New Delhi 110012, India; chowdary_isro@yahoo.com
4 Department of Geography, Virginia Tech, Blacksburg, VA 24061, USA; rsantosh@vt.edu
* Correspondence: monidipadas@hotmail.com

Citation: Das, M.; Ghosh, S.K.; Chowdary, V.M.; Mitra, P.; Rijal, S. Statistical and Machine Learning Models for Remote Sensing Data Mining—Recent Advancements. *Remote Sens.* **2022**, *14*, 1906. https://doi.org/10.3390/rs14081906

Received: 24 March 2022
Accepted: 7 April 2022
Published: 15 April 2022

During the last few decades, the remarkable progress in the field of satellite remote sensing (RS) technology has enabled us to capture coarse, moderate to high-resolution earth imagery on weekly, daily, and even hourly intervals. This wealth of earth surface data, if analyzed effectively, can provide significant insights into various geo-spatial processes, and eventually, can help us in making crucial decisions in a timely manner. Nevertheless, these RS data, as continuously captured at varying spatial, spectral, and temporal resolutions, are not only voluminous but also acquired heterogeneous data, where diverse categories of sensors, i.e., optical/microwave were used. Consequently, mining useful patterns/information from these enormous volumes of heterogeneous unstructured data requires enhanced data mining techniques exploiting the power of advanced computational intelligence and high-performance computing paradigms. Moreover, in the context of resolving urgent issues, such as in environmental nowcasting, a timely analysis of the RS data requires resource-efficient computation models with real-time processing and online analysis capabilities [1,2].

With this background in mind, in this Special Issue, we called for high-quality papers focusing on recent advancements in conventional statistical as well as machine learning techniques and modern AI (artificial intelligence)-driven technologies for efficient mining of remote sensing data. This Special Issue also aimed to provide a common platform for professionals, researchers, and practitioners from heterogeneous communities, including artificial intelligence, machine learning, geographical information systems, and spatial data science, to share their views, innovations, research achievements, and solutions to foster the advancement of intelligent analytics and efficient management of remote sensing data. Papers were invited to cover the following broad topics:

- Advanced and energy-efficient machine learning models for RS data mining
- Enhanced statistical and scalable computing methods for RS data mining
- Real-time processing and online analytics of RS data
- Real-world applications of RS data mining

After the rigorous review process, a total of five papers have been accepted for publication in this issue. The selected papers either deal with the core challenges, such as missing data handling, noisy label distillation, feature-level fusion, etc., in remote sensing data analyses [3–5], or these highlight on various critical real-world problems, including oil spill detection [6], and high wind speed inversion [7].

As just mentioned above, missing data is a common problem in the field of remote sensing data analytics. This primarily occurs due to internal malfunctioning of the satellite remote sensing devices/sensors or due to the poor atmospheric condition, such as the presence of thick cloud cover. Remote sensing images with missing information not only reduce the usability of the data but also may negatively affect the performance of the analytical models. The problem appears to be more prominent at the time of analyzing aerosol optical depth (AOD) from remotely sensed data. AOD is a key parameter reflecting the characteristics of aerosols, and plays a significant role in predicting the concentration of pollutants in the atmosphere. However, as highlighted in the work of Chi et al. [3], the AOD data obtained by satellites are often found to be missing, and thereby, impose serious research challenges. The existing methods of AOD recovery primarily focus on to the accuracy of AOD restoration while neglecting the AOD recovery ratio. In order to solve the issue, Chi et al. [3] have proposed a light gradient boosting-based two-step model, termed as TWS, that fills the missing AOD data by combining data from multiple sources and at the same time learning spatio-temporal relationships of AOD. Experimental evaluation of TWS with respect to recovering AOD from Terra Satellite's 2018 AOD product has demonstrated the reliability of TWS method in producing competitive and consistent performance in AOD restoration. Overall, the work of Chi et al. [3], as included in our Special Issue, is of great significance in the context of studying the spatial distribution of atmospheric pollutants and handling missing data in this context.

In spite of the huge availability of remotely sensed data in recent years, the data are often found to be annotated with noisy labels. Label noise occurs whenever there is a mismatch between the ground truth label and the observed label. This happens due to several reasons, including manual labeling error, wrong or misinterpretation of the data, and so on. Noisy label can lead to serious network over-fitting problem and may negatively impact on the model performance. Therefore, noisy label distillation plays an important role in remote sensing image scene classification or segmentation tasks. Traditional models are typically based on direct fine-tuning and pseudo-labeling approaches, which are not only inefficient but also, may badly influence the model in other ways. In order to address such problem, in this Special Issue, Zhang et al. [4] have proposed a novel noisy label distillation approach grounded on an end-to-end teacher-student framework, which does not require pre-training on clean or noisy data. Evaluation on benchmark remote sensing image datasets with injected noise has demonstrated the superiority of the proposed approach [4] over the state-of-the-art techniques.

Apart from dealing with the core challenges in remote sensing data analytics, another way of improving the model performance is to fully exploit the increasingly sophisticated data from multiple sources. For example, the optical remote sensing data provides us with significantly larger amount of spectral information compared to the images captured using synthetic aperture radar (SAR), whereas the SAR technology has more penetration capability and has the advantage of generating images almost in all weather conditions. Remote sensing image fusion is, thus, important for enhancing the application ability of remote-sensing images, and accordingly, it has gained immense research attention in recent years. Incidentally, the remote-sensing image fusion can be performed both at the pixel-level and at the feature-level. However, in contrast with the pixel-level fusion, feature-level fusion considers more diverse factors, and thereby, helps to obtain more macro-level information than that obtained using pixel-level fusion. This Special Issue includes an interesting article by Kong et al. [5] on feature-level fusion-based classification of remote sensing images using features extracted from polarized SAR and optical images. The approach is based on a combination of Random Forest (RF) and Conditional Random Fields (CRFs). Typically, the model exploits the power of CRF in spatial context feature modeling and improves the RF-based classification. Experimental evaluation shows the efficacy of the proposed fusion-based classification approach.

In addition to discussion on the technical challenges being faced during remote sensing image data processing, this Special Issue also includes papers on some critical applications of remote sensing data analytics [6,7]. For example, in the fourth article of this Special Issue, Almulihi et al. [6] have presented the application of SAR Image analysis in oil spill detection. The proposed approach is based on online extended variational learning of dirichlet process mixtures of Gamma distributions. The technical novelty lies here in extending the finite Gamma mixture model that can handle infinite number of mixture components. The online learning property of the proposed model makes it more advantageous over the batch learning-based models at the time of dealing with massive and streaming data. Empirical study with respect to real-world application of oil spill detection from SAR images demonstrates the effectiveness of the approach proposed by Almulihi et al. [6].

High wind speed inversion is another critical as well as challenging application of remote sensing data analytics, which has gained significant research interest in present days. Wind speed is one of the key sea surface parameters that prominently influence diverse oceanic applications. The traditional ways of detecting wind speed using remote sensing imaging technology are often found to be failed when the wind speed is high. The study made by Zhang et al. [7], as included in this Special Issue, reveals that machine learning techniques can be effectively employed as the complements of these conventional RS technology-based models. Experimentations on multi-sourced RS data show that machine learning schemes of Support Vector Regression (SVR), combined Principal Component Analysis (PCA) and SVR (PCA-SVR), and Convolutional Neural Network (CNN) can be certainly useful for improving the accuracy in high wind speed inversion on sea surface, where CNNs are promising models in this area.

We hope that the readers will become highly benefitted from the insightful discussions and presentations of our Special Issue papers, as concisely discussed above, and also will be encouraged to contribute to these rapidly progressing areas.

Funding: This research received no external funding.

Acknowledgments: We would like to thank all authors who have contributed to this volume by sharing their domain knowledge, research experiences and experimental results.

Conflicts of Interest: The authors declare no conflict of interest.

References

1. Das, M. Real-time prediction of spatial raster time series: A context-aware autonomous learning model. *J. Real-Time Image Process.* **2021**, *18*, 1591–1605. [CrossRef]
2. Das, M. Online prediction of derived remote sensing image time series: An autonomous machine learning approach. In Proceedings of the IGARSS 2020—2020 IEEE International Geoscience and Remote Sensing Symposium, Waikoloa, HI, USA, 26 September–2 October 2020; pp. 1496–1499.
3. Chi, Y.; Wu, Z.; Liao, K.; Ren, Y. Handling missing data in large-scale MODIS AOD products using a two-step model. *Remote Sens.* **2020**, *12*, 3786. [CrossRef]
4. Zhang, R.; Chen, Z.; Zhang, S.; Song, F.; Zhang, G.; Zhou, Q.; Lei, T. Remote sensing image scene classification with noisy label distillation. *Remote Sens.* **2020**, *12*, 2376. [CrossRef]
5. Kong, Y.; Yan, B.; Liu, Y.; Leung, H.; Peng, X. Feature-Level Fusion of Polarized SAR and Optical Images Based on Random Forest and Conditional Random Fields. *Remote Sens.* **2021**, *13*, 1323. [CrossRef]
6. Almulihi, A.; Alharithi, F.; Bourouis, S.; Alroobaea, R.; Pawar, Y.; Bouguila, N. Oil spill detection in SAR images using online extended variational learning of dirichlet process mixtures of gamma distributions. *Remote Sens.* **2021**, *13*, 2991. [CrossRef]
7. Zhang, Y.; Yin, J.; Yang, S.; Meng, W.; Han, Y.; Yan, Z. High Wind Speed Inversion Model of CYGNSS Sea Surface Data Based on Machine Learning. *Remote Sens.* **2021**, *13*, 3324. [CrossRef]

Article

Handling Missing Data in Large-Scale MODIS AOD Products Using a Two-Step Model

Yufeng Chi [1,2], Zhifeng Wu [1,2], Kuo Liao [3] and Yin Ren [1,2,*]

1 Key Laboratory of Urban Environment and Health, Fujian Key Laboratory of Watershed Ecology, Institute of Urban Environment, Chinese Academy of Sciences, Xiamen 361021, China; yfchi@iue.ac.cn (Y.C.); zfwu@iue.ac.cn (Z.W.)
2 University of Chinese Academy of Sciences, Beijing 100049, China
3 Wuyishan National Climatological Observatory, Wuyishan 354200, China; liaok@cma.gov.cn
* Correspondence: yren@iue.ac.cn; Tel.: +86-0592-619-0697

Received: 23 October 2020; Accepted: 16 November 2020; Published: 18 November 2020

Abstract: Aerosol optical depth (AOD) is a key parameter that reflects the characteristics of aerosols, and is of great help in predicting the concentration of pollutants in the atmosphere. At present, remote sensing inversion has become an important method for obtaining the AOD on a large scale. However, AOD data acquired by satellites are often missing, and this has gradually become a popular topic. In recent years, a large number of AOD recovery algorithms have been proposed. Many AOD recovery methods are not application-oriented. These methods focus mainly on to the accuracy of AOD recovery and neglect the AOD recovery ratio. As a result, the AOD recovery accuracy and recovery ratio cannot be balanced. To solve these problems, a two-step model (TWS) that combines multisource AOD data and AOD spatiotemporal relationships is proposed. We used the light gradient boosting (LightGBM) model under the framework of the gradient boosting machine (GBM) to fit the multisource AOD data to fill in the missing AOD between data sources. Spatial interpolation and spatiotemporal interpolation methods are limited by buffer factors. We recovered the missing AOD in a moving window. We used TWS to recover AOD from Terra Satellite's 2018 AOD product (MOD AOD). The results show that the MOD AOD, after a 3 × 3 moving window TWS recovery, was closely related to the AOD of the Aerosol Robotic Network (AERONET) (R = 0.87, RMSE = 0.23). In addition, the MOD AOD missing rate after a 3 × 3 window TWS recovery was greatly reduced (from 0.88 to 0.1). In addition, the spatial distribution characteristics of the monthly and annual averages of the recovered MOD AOD were consistent with the original MOD AOD. The results show that TWS is reliable. This study provides a new method for the restoration of MOD AOD, and is of great significance for studying the spatial distribution of atmospheric pollutants.

Keywords: LightGBM; spatiotemporal weight interpolation; AOD recovery; East Asia

---

1. Introduction

Atmospheric aerosols are a dispersion system of suspended colloids formed by solid or small particles [1]. With the increase in the number of aerosols emitted by human activities, the scattering and absorption of solar radiation forms a brighter cloud layer and directly reduces the efficiency of precipitation [2]. Moreover, the increased number of aerosols changes the structure of the atmosphere, reduces solar radiation on the surface, increases the heating effect on the atmosphere, reduces precipitation, and inhibits the removal of pollutants [3]. Additionally, the weak water cycle brought about by aerosols directly affects the quality and quantity of fresh water [4]. Therefore, it is crucial to quantitatively measure the aerosol optical depth (AOD). Typically, the definition of AOD is the vertical integral of the aerosol extinction coefficient in the atmosphere column, which is used to describe the aerosol optical properties [5,6].

There are two main methods for obtaining AOD data: ground acquisition and space acquisition. The Aerosol Robot Network (AERONET) represents the ground observation network, which relies mainly on a sun spectrophotometer to conduct fully automatic measurements of the AOD at the instrument deployment site [7,8]. Compared with space acquisition, the AOD obtained by the ground observation network has higher accuracy. Nevertheless, it is difficult to provide a wide range of viewing angles for the AOD of ground measurements due to limitations in equipment deployment and observation ranges [9,10]. Therefore, it is more efficient to use remote sensing for AOD measurement and inversion on a large scale. The following are some of the remote sensing inversion products that are commonly used in AOD observations: (1) MODIS sensors on the Terra and Aqua satellites in polar orbit are used to provide global AOD products (MOD AOD and MYD AOD) with resolutions of 10 km and 3 km every day through the Dark Target (DT) [11] and Dark Blue (DB) [12] algorithms [13,14]. (2) MODIS sensors combined with the Multi-Angle Implementation of Atmospheric Correction (MAIAC) algorithm [15] are used to provide AOD products with a fixed 1-km grid. MAIAC AOD uses time series to detect multiangle surface features to recover Bidirectional Reflectance Distribution Function (BRDF). Compared with the DT and DB algorithm, it can better identify AOD information in cloud and snow areas [16,17]. (3) The Advanced Himawari Imager (AHI) sensor on the Japan Himawari-8 geostationary satellite provides AOD products with a spatial resolution of 5 km at a spectral wavelength of 500 nm and continuously monitors East Asia at a maximum interval of 10 min [18,19].

At present, many studies use AOD as an important indicator or parameter for the mapping of air pollutants (e.g., PM2.5, PM10) [20–22]. Complete and high-precision AOD distribution data will greatly improve the quality of the mapping of air pollutants. However, uncertainties in cloud detection, limitations of the AOD inversion algorithm, and sensor degradation are the three main factors that cause a partial loss of the AOD local data retrieved by satellites [23–25]. For example, the shortcomings of the DT algorithm and DB algorithm for AOD detection in bright areas, the errors of cloud detection in some heavily polluted areas and the degradation of other sensors directly affect the detection of dark pixels in low angle areas, which leads to the loss of AOD data in some areas [26,27]. A study of the Yangtze River Delta in China found that the missing rate of MOD AOD reached 89.6% between 2014 and 2017 [28]. Because the results of AOD are affected by meteorological conditions, human activities and vegetation coverage, it is difficult to ensure the accuracy of the AOD restoration [29].

A large quantity of research has focused on how to recover missing information from AOD data. One approach is through the innovation of the inversion algorithm to reduce the missing AOD. For example, some researchers use low cloud detection standards or the Dense Dark Vegetation (DDV) algorithm to improve the AOD inversion accuracy of bright surfaces [30,31]. However, such methods still cannot overcome the missing AOD data caused by cloud shading [32]. Statistical regression models such as linear regression [33,34], spatial interpolation and spatiotemporal interpolation [35,36] are used to fill in the deficiency of the AOD statistical regression models, and it is difficult to analyze the internal relationships of the global heterogeneity of the AOD data, which results in poor recovery results. AOD information is filled in by using a machine learning methods such as random forest (RF) [20] or gradient boosting machine (GBM) [24] to process the multisource data. The strong data mining ability of the machine learning methods is good for fitting multisource data, and it can achieve higher precision at the same time [9,37].

In this paper, a two-step model (TWS) is proposed to recover the missing AOD caused by the presence of clouds of MOD AOD under the premise of ensuring recovery accuracy. Specifically, the first step of TWS uses a machine learning method to integrate multisource AOD data. The second step uses the spatio-temporal interpolation and spatial interpolation methods of moving windows to further fill in the missing MOD AOD. In addition, the second step of TWS uses a local buffer to reduce the heterogeneity of the AOD caused by time differences. Section 2 of this paper describes the research area and data set, Section 3 shows the methodology of the TWS, Section 4 shows the results of the model, Section 5 discusses the model and application, and Section 6 presents the conclusions.

2. Materials

2.1. Study Areas

Part of the East Asia region (18–50° N, 96–150° E) was selected as the study area (Figure 1). The research area mainly includes regions of China, Mongolia, Japan, the Korean Peninsula, and the Northeast Pacific. The study area includes countries that contain more than 75% of the population distribution in East Asia in total (central and eastern China, Korean Peninsula, Japan, Mongolia, northern Vietnam) and major urban agglomerations (Yangtze River Delta, Pearl River Delta, Seoul City Cluster, Tokyo City Cluster) [38]. The spatial and temporal distribution characteristics of AOD data are complicated by the increasing number of human activities [39]. Additionally, a large-scale research area can reduce the probability of all missing AOD data on a given day and provide enough data for research. Moreover, a larger study area has more complex land types and other factors, which can better test the universality of the model.

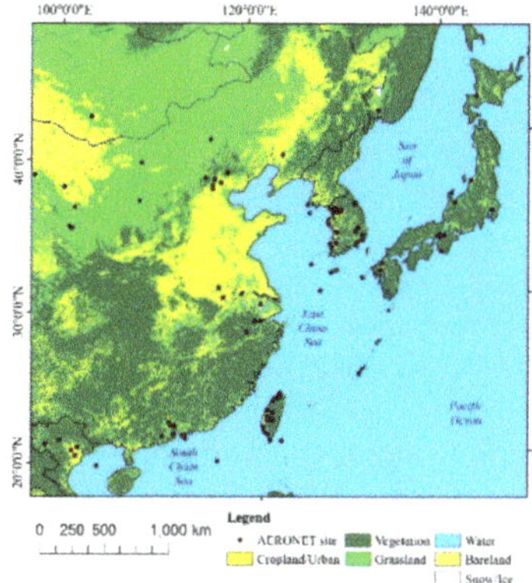

Figure 1. Distribution of the AERONET sites considered in this paper.

2.2. Datasets

We collected the data from 86 ground AERONET stations in the study area from 31 December 2017, to 1 January 2019 (Figure 1) and the satellite AOD dataset. The satellite data included Terra and Aqua satellite AOD products (MOD AOD/MYD AOD), MAIAC AOD, and AHI AOD products. In addition, we included part of the auxiliary data.

2.2.1. AOD Products

We selected the following three AOD products: 1. The "MOD AOD" data were selected from MODIS Terra, and the "MYD AOD" data were from Aqua Aerosol Collection 6.1, which were downloaded through Earthdata (https://earthdata.nasa.gov). A total of 16,233 images of MOD AOD and MYD AOD were selected with a time resolution of one day and the spatial resolution of 3 km [40]. 2. More than 19508 MAIAC AOD data were downloaded from Earthdata. We selected the MAIAC AOD data at the spectral wavelength of 550 nm and then removed invalid AOD based on the guidance of the filter quality assurance in the user manual (reserve AOD when QA.CloudMask = Clear and QA.AdjacencyMask = Clear). 3. We selected the Advanced Himawari-8 AOD (AHI AOD), which is provided by the Japan Meteorological Agency (JMA). AHI AOD data were divided into two levels:

L2 and L3. The L3 product selected in this research underwent strict cloud screening. Therefore, the L3 product has higher accuracy and reliability than L2 [41]. L3 daily products (averaged from L3 hour products) have a spatial resolution of 5 km and contain a total of 367 images. AHI AOD date were obtained from the FTP provided by JMA (ftp.ptree.jaxa.jp).

2.2.2. AERONET Data

AERONET (aeronet.gsfc.nasa.gov) has a time resolution of 15 min. AERONET AOD contains three quality levels: Level 1.0 (unscreened), Level 1.5 (cloud-screened and quality controlled), and Level 2.0 (quality-assured). Compared with Level 1.0, the uncertainty of Level 1.5 and Level 2.0 in version 3 is low [8]. In this paper, the Level 1.5 and Level 2.0 data of version 3 of the AERONET site in 86 research areas are used as ground truth values for comparison.

2.2.3. Auxiliary Data

The auxiliary data were mainly divided into meteorological, terrain, land data and other types. The meteorological data were extracted from the Modern-Era Retrospective analysis for Research and Applications version 2 (MERRA2) dataset (https://earthdata.nasa.gov) [42]. The meteorological data included the temperature (TLML), wind speed (WS), surface roughness (ZM), surface specific humidity (QSH), and planetary boundary layer height (PBLH). The spatial resolution of the meteorological data was $0.625° \times 0.5°$, and the average value of the 9:00–12:00 local time (satellite transit time) data was calculated as the meteorological data of the day. The terrain data were extracted from Shuttle Radar Topography Mission (SRTM) data (https://earthdata.nasa.gov) with a spatial resolution of 90 m. The terrain data included the digital elevation model (DEM), slope, and aspect. The land data included population data, road density, and Normalized difference vegetation index (NDVI) composition. The population data were obtained by LandScan (landscan.ornl.gov), which is integrated by multisource data and released once per year. The spatial resolution of the population data was approximately 1 km [43]. The road data provided by OpenStreet (www.openstreetmap.org) were mainly composed of data shared by users, and were therefore free from copyright. The road data were the vector data format of ESRI (RL). NDVI data use MOD13 A2 16D 1 km spatial resolution (collection 6) data (https://earthdata.nasa.gov) [44]. Other types included the day of the year (DOY).

3. Methods

Due to aerosol diffusion, AOD inversion algorithm differences, remote sensing image detection time differences, and differences in multisource AOD data are mainly reflected in the different data sources, different data detection times, and various data detection positions [10,45,46]. Thus, the life cycle of aerosols in the troposphere varies from a few days to a few weeks [4,47]. Over a short time, there is a correlation between different AOD data sources; in addition, there is a correlation between different AOD data detection times. According to the 2018 statistics of the AOD data in the study area, the MOD AOD at the same location on the same day is directly related to MYD AOD, MAIAC AOD and AHI AOD data. The MOD AOD at the same position correlates with that of the adjacent time, and the specific data are shown in Table 1. The spatial correlation refers to the correlation coefficient (R) of the effective AOD values of two adjacent pixels. The time correlation refers to the R of the effective value of the target AOD pixel and the adjacent day AOD pixel.

Table 1. MOD AOD correlation (spatial correlation, temporal correlation, and correlation of different AOD data sources).

	R		
MOD AOD spatial correlation **MOD AOD time correlation**		R = 0.92 (n = 13,489,645) R = 0.57 (n = 15,895,438)	
Time correlation of multisource AOD data (compared with MOD AOD)	MYD	MAIAC	AHI
	R = 0.56 (n = 7,746,528)	R = 0.77 (n = 10,125,868)	R = 0.56 (n = 15,256,795)

Note: n represents the number of observations.

This paper proposes a two-step model (TWS) that combines the rich data volume of multisource data and the inherent spatial-temporal distribution relationships of aerosols to recover missing MOD AOD. First, we preprocess the multisource data and then use the TWS method to recover the MOD AOD. 1. For the multisource AOD data obtained at the same spatial location on the same day, some sources have pixel values, and some are missing. The existing data helps to recover some of the missing MOD AOD values from the other data sources, which is possible due to the complementarity of the multisource AOD data. The multisource AOD data is fitted and calculated using a machine learning method, and then the overlapping parts of the multisource AOD data are calculated by a weighted average to fill in some missing MOD AOD pixels. 2. In the moving window, the missing MOD AOD can be recovered through space or spatiotemporal relationships. First, we create a moving window. The corresponding calculation scenario is determined by the number and distribution of the AOD in the moving window and then combined with the buffer factor to perform the calculation. Finally, the recovered MOD AOD pixels are obtained by the priority settings of the overlapped pixels (priority stack). The steps of the specific method are shown in Figure 2.

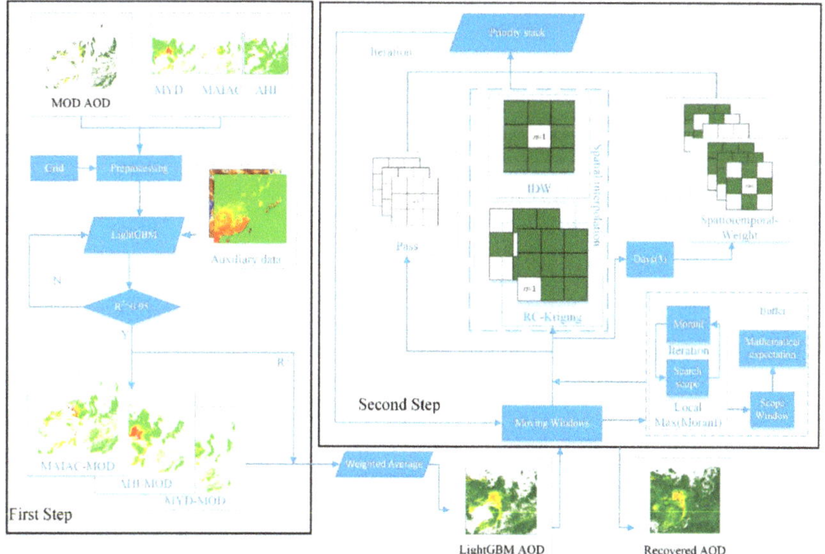

Figure 2. Flowchart of the proposed TWS model. The recovered AOD represents the final result.

3.1. Data Preprocessing

First, we create a 3-km spatial resolution grid in the UTM coordinate system. We rebuild the multisource data according to the grid position (including the AOD data set and auxiliary data). MAIAC AOD, AHI AOD, meteorological data, terrain data, and land data must be reconstructed because the spatial resolution is not 3 km. Specifically, MAIAC AOD, terrain data and NDVI must have their averages calculated in the 3-km grid. We summarize the population data within the 3-km grid (POP), and the RL data must have the total length of the roads in the grid calculated, which is assigned to the road length grid (RLG). All of the reproduced information must be resampled due to pixel position deviation.

3.2. First Step of TWS

GBM uses a gradient descent algorithm to adjust the regression tree of the weak learner's addiction model, thereby reducing the loss of the objective function. LightGBM was developed by Microsoft and uses the GBM framework. LightGBM adds Gradient-based One-Side Sampling (GOSS) and Exclusive Feature Bundling (EFB). Compared with GBM, LightGBM can accelerate the calculation speed under the premise of ensuring accuracy, and has a higher calculation speed for large sample data [48,49]. In this study, MOD AOD was used as the dependent variable; MYD AOD, MAIAC AOD, AHI AOD and the other auxiliary data were used as independent variables. Three LightGBM models, i.e., MOD-MYD, MOD-MAIAC and MOD-AHI, were established. Then, the accuracy of the prediction model was verified by a 10-fold cross-validation method. The data for constructing the LightGBM model were randomly divided into ten groups. Cyclic verification was performed ten times, and one group was used for prediction verification, while the remaining nine were used as training samples. The decision coefficient (R^2) was used as an index for model verification. Next, we used the trained model to predict the missing AOD of MOD AOD where MYD AOD, MAIAC AOD, and AHI AOD were not missing. After calculating the three LightGBM models, weighted average processing was performed on the overlapping pixels according to the LightGBM training result R^2.

3.3. Second Step of TWS

AOD data has a strong spatial correlation (the R of adjacent MOD AOD is 0.9), but it also has a certain correlation in time (the R of adjacent time MOD AOD is 0.5). Therefore, when restoring MOD AOD information, we consider the spatial relationship of AOD and the spatiotemporal relationship. Moreover, the small moving window could reduce the uncertainty caused by AOD spatial heterogeneity.

3.3.1. Design of Moving Window Size and Selection of Interpolation Mode

Moving windows of different sizes will affect the number of valid MOD AOD pixels. However, a large moving window will cause serious spatial heterogeneity of MOD AOD, and will also affect the computing performance of the MOD AOD recovery. In this study, we set the size of the moving window to 3 × 3 pixels, 7 × 7 pixels, and a self-adaption window (from 3 pixels to 7 pixels) [34]. The self-adaption window is determined by the ratio of the number of valid AOD pixels to the total number of pixels. The formula is as follows:

$$Sw = Max\left(\frac{PV_x}{PA}\right) \quad x \in (3,4,5,6,7) \tag{1}$$

where Sw represents the size of the self-adaption window; PV_x is the number of valid AOD pixels in the window; and PA is the total number of pixels in the window.

Spatial interpolation and spatiotemporal interpolation methods have good adaptability to recover the AOD data, which performs a strong correlation in local space and is spatiotemporal. Regardless of whether it is spatial interpolation or spatiotemporal interpolation, the recovery results of the AOD data are greatly affected by the distribution and the number of valid AOD data points and the spatiotemporal distribution of the AOD data. Therefore, this study designed the following scenarios (taking a 3 × 3 window as an example), as shown in Figure 3: (1). Inverse Distance Weight interpolation (IDW) [50] is a spatial interpolation method. It was applied when the central MOD AOD was missing in the moving window. (2). We used region constraints kriging (RC kriging) which involves adding a constraints factor to the ordinary kriging method. It was applied when five or fewer pixels of MOD AOD were missing in the moving window. (3). We used spatiotemporal weight interpolation when the number of missing cells of Day 2 MOD AOD was greater than or equal to 5 and the number of valid AOD cells of Day 1 or Day 3 MOD AOD was greater than or equal to 5. (4). When there were too few MOD AOD pixels in the moving window for three consecutive days (Day 2 had no MOD AOD pixels and the number of valid MOD AOD cells for Days 1 and 3 were fewer than (5), we ignored this part of the calculation. The change in the window size changed the above rule (the ratio of the number of AOD pixels to the total number of moving window pixels). For example, when the window was 7 × 7, the five pixels in condition two increased to 27.

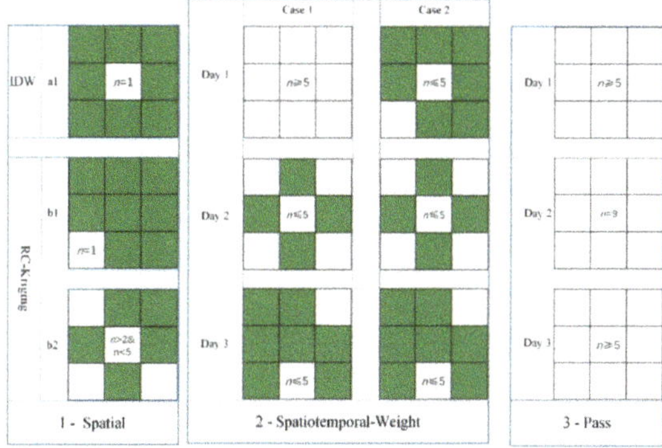

Figure 3. Three scenarios of the second step TWS. Here, n represents the number of missing AOD pixels in the moving window, and Days 1, 2, and 3 represent three consecutive days (where Days 1 and 3 are disordered). 1—Spatial represents spatial interpolation, including IDW and RC kriging. 2—Spatiotemporal-weight represents spatiotemporal weighted interpolation and lists two examples. 3—Pass indicates that this scenario ignores and does not calculate the AOD in the moving window.

3.3.2. Buffer Factor

Because the moving window introduced only a small quantity of MOD AOD data, it caused the prediction value to deviate greatly between the spatial interpolation and spatiotemporal interpolation of MOD AOD. Therefore, a buffer factor was introduced to correct the deviation. Global Moran's I (MoranI) [51] is a statistic for spatial autocorrelation; the larger the MoranI of AOD, the higher the similarity of the AOD data, which can provide more information for the recovery of AOD gaps. This approach is applied to calculate the spatial autocorrelation of MOD AOD in the region; the larger the value of MoranI, the higher the correlation of the MOD AOD data in the region. This study calculated MoranI in different areas and determined the maximum amount of MoranI in a local area. The corresponding local area was called the scope window (Figure 4). The mathematical expectation of the MOD AOD of the scope window served as a buffer factor for the spatial interpolation of the MOD AOD. Uncertainty in the numeric values of the MOD AOD pixels in the scope window was prone to occur, and the MOD AOD pixel values were not in a Gaussian distribution. The Spearman correlation coefficient was introduced as the time buffer factor of the MOD AOD. The mathematical expectation of the Spearman correlation coefficient for three consecutive days and the MOD AOD of the scope window were used as buffer factors for the spatiotemporal interpolation of the MOD AOD. The formula is as follows:

$$\begin{gathered} \text{MoranI} = \frac{n\sum_{i=1}^{n}\sum_{j=1}^{n}G_{ij}(p_i-\overline{p})(p_j-\overline{p})}{\sum_{i=1}^{n}\sum_{j=1}^{n}G_{ij}\sum_{i=1}^{n}(p_i-\overline{p})^2} \\ G_{ij} = 1/\sqrt{(i_x-j_x)^2+(i_y-j_y)^2} \\ w \leftarrow \textit{Scope Window} \leftrightarrow \textit{Max}(\text{MoranI}_{w-1}, \text{MoranI}_w, \text{MoranI}_{w+1}) \\ E_w = \left(\sum_{i=1}^{w*w} S_i\right)/w^2 \\ P_{(S_{tk},E_{t2})} = \frac{\sum_{j=1}^{n}(S_{tk}-E_{t2w})(S_{tk}-\overline{\tau_{t2}})}{\sqrt{\sum_{j=1}^{n}(S_{tk}-\overline{\tau_{tk}})^2\sum_{j=1}^{n}(S_{tk}-\overline{\tau_{t2}})^2}} k \in (1,3) \end{gathered} \quad (2)$$

where MoranI represents the Global Moran's I. Here, n represents the number of valid pixel AODs; p_i and p_j represent the AOD values of the two pixels, I and J; $\overline{x}$ represents the average value of the AOD pixels; $dis(i,j)$ represents the spatial distance between the two pixels, I and J; $G_{i,j}$ represents the inverse distance weight; *Scope Window* represents the window that corresponds to the maximum local MoranI, *Scope Window* is a square; w represents the number of pixels on one side of the square a *Scope Window*; $\leftrightarrow$ represents iterative search for the *Scope Window*; $\leftarrow$ represents obtaining w; S_i represents the AOD value in the *Scope Window*; S_{tk} represents the AOD value in the *Scope Window* on day *tk*; E_w represents the mathematical expectation of AOD in the *Scope Window* (buffer factor); and $P_{(S_{tk},E_{t2})}$ represents the Spearman correlation coefficient between day *tk* and day *t2*.

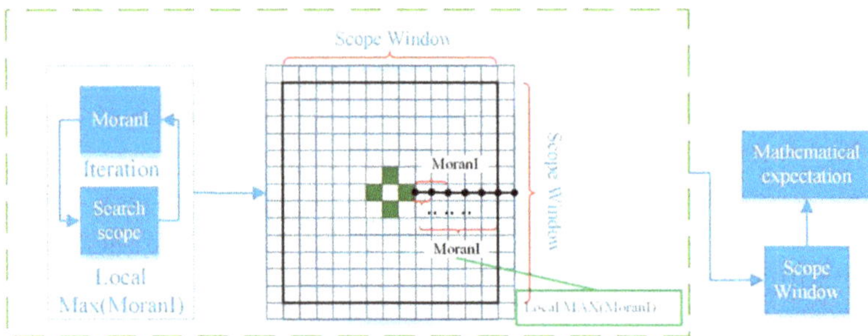

Figure 4. Buffer factor calculation flowchart.

3.3.3. Spatial Interpolation Method (IDW and RC Kriging)

Compared with other complicated physical models of AOD recovery, the spatial interpolation of AOD can quantify the spatial information of the AOD with known spatial positions, which can easily and effectively predict the missing AOD data over a small range. Moreover, the AOD spatial interpolation method does not require an excessive number of parameters. Among them, IDW and the spatial interpolation method are commonly used to predict the missing AOD. Additionally, based on the best linear unbiased prediction of ordinary kriging interpolation [52], we introduced the buffer factor for spatial interpolation when predicting the MOD AOD in a moving window, and established

RC kriging. The buffer factor helps the RC kriging method to better adapt to the stability of mod AOD in the moving window [53]. The formula is as follows:

$$Z_1 = \left[\sum_{i=1}^{N}\sum_{j=1}^{N} G_{i,j}\left(S_{i,j} - E_w\right)\right] + E_w \begin{cases} \sum_{i=1}^{N}\sum_{j=1}^{N} \daleth 2_{i,j} \times Cov\left(s_{i,j}\right) - \mu = Cov\left(s_{j,i}\right) \\ \sum_{i=1}^{N}\sum_{j=1}^{N} \daleth 2_{i,j} = 1 \\ Z_2 = \left[\sum_{i=1}^{N}\sum_{j=1}^{N} \daleth 2_{i,j}\left(S_{i,j} - E_w\right)\right] + E_w \end{cases} \quad (3)$$

where Z_1 and Z_2 represent the AOD estimates produced by IDW interpolation and RC Kriging interpolation, $G_{i,j}$ represents the inverse distance weight, $s_{i,j}$ represents the MOD AOD value at points I and J, μ represents the Lagrange multiplier, $\daleth 2_{i,j}$ represents the weight, $Cov\left(s_{i,j}\right)$ and $Cov\left(s_{j,i}\right)$ represent the covariance of $s_{i,j}$ and $s_{j,i}$, and E_w represents the mathematical expectation in the *Scope Window* (buffer factor).

3.3.4. Spatiotemporal Weight Interpolation (STW)

Spatiotemporal interpolation can effectively consider both space and time MOD AOD relationships and overcome the shortcomings of MOD AOD space interpolation [54]. We quantify the time distance of one day of MOD AOD as 1 and combine the spatial distance between the MOD AOD pixels to determine the spatiotemporal distance. The spatiotemporal distance and the buffer factor are used to determine the spatiotemporal weight of MOD AOD spatiotemporal interpolation. We combine the spatiotemporal interpolation and spatiotemporal weights to generate spatiotemporal weight interpolation (STW). In this study, the time of STW used for MOD AOD was set to three days (including the predicted day, as well as the days before and after the predicted time), to avoid the excessive AOD data noise caused by a time span that is too long. The specific formula is as follows:

$$\begin{gathered} Z_{ST_0} = \sum_{tk=1}^{3}\left(\sum_{j=1}^{N_t}\left(\left[\sum_{i=1}^{N_t}\left(\daleth_{tk_{i,j}}\left(S_{t_{i,j}} - E_{tw}\right)\right)\right] + E_{tw}\right)\right) \\ \daleth_{tk} = \daleth_{(tk,tk)} = \sum_{j=1}^{N}\sqrt{\left(1 - \left[\left(P_{(S_{t1},E_{tk})} + P_{(S_{tk},E_{t3})}\right)/2\right]\right)\left(\frac{1/dis(tk_i,tk_j)}{\sum_{i=1}^{N}\left(1/dis(tk_i,tk_j)\right)}\right)} \quad k = 2 \\ \daleth_{tk} = \daleth_{(tk,t2)} = \sum_{j=1}^{N}\sqrt{\left(\frac{P_{(S_{tk},E_{t2})}}{2}\right)^2 + \left(\frac{1}{dis\left(tk_i,tk_j\right)} / \sum_{i=1}^{N}\left(1/dis\left(tk_i,tk_j\right)\right)\right)^2} \quad k \in (1,3) \\ dis(i,j) = \sqrt{\left(i_x - j_x\right)^2 + \left(i_y - j_y\right)^2} \end{gathered} \quad (4)$$

where Z_{ST_0} represents the estimation of STW. T represents the time of day, t1 is the previous day, t2 is the day to be calculated, and t3 is the next day. S_t represents the value of the valid AOD. E_{tw} is the mathematical expectation in *Scope Window* within t days (buffer factor), $P_{(\tau_{tk},\tau_{t2})}$ represents the R between *t*2 and *tk*. $\daleth_{tk}$ represents the time weight of k days ($k \in (1,2,3)$). N is the number of pixels in the moving window, and $dis\left(tk_i, tk_j\right)$ represents the spatial distance between tk_i and tk_j.

3.3.5. Priority Setting of Overlapping Pixels

Because the spatial interpolation of MOD AOD and STW belong to the second step in TWS, TWS will have overlapping results of MOD AOD recovery with the movement of the window. Therefore, TWS should set the priority of the MOD AOD recovery results. The priority of the MOD AOD recovery result was set to IDW > RC Kriging > STW. If the MOD AOD recovery resulted in overlap, then the missing values of the MOD AOD were filled according to their priority. Furthermore, if the recovery results of the MOD AOD overlap in the same method, the average amount of the MOD AOD recovery results overlap should be calculated as the final result of the MOD AOD. For example, in the process of moving the window of TWS, RC kriging and STW were used in the two calculations before and after the predicted time, and the overlapping area of the MOD AOD recovery result should have used the RC kriging result. If RC kriging was used for both calculations during the window movement of the TWS, the overlapping area of the MOD AOD recovery results were calculated in the average value as the final MOD AOD recovery result.

3.3.6. Validation Methodology

A comparison between the MOD AOD recovery results and AERONET data can be used as the basis for the MOD AOD recovery accuracy [55]. The time resolutions of MOD AOD and AERONET were different. This research calculated the transit time of the satellite (Terra) (30 min before and after) and compared the average value of the AERONET data with the MOD AOD data of the location pixels for the AERONET site [37]. In addition, AERONET collected AOD data of multiple wavelengths, many of which were slightly different from the MOD AOD wavelength (550nm). Therefore, the AERONET AOD at 550 nm was interpolated using the Ångström exponent [7]. In addition, both the 551 nm and 560 nm AOD data were used in the AERONET data to evaluate the MOD AOD. The specific calculation formula is as follows:

$$\tau_{\omega} = \beta\omega^{\delta}$$
$$\delta = -\frac{\ln(\tau_1/\tau_2)}{\ln(\omega_1/\omega_2)} \tag{5}$$
$$\beta = \tau_1(\omega_1)^{\delta} = \tau_2(\omega_2)^{\delta}$$

where τ, τ_1, and τ_2 represent the AOD at wavelengths ω, ω_1, *and* ω_2, respectively. Here, δ represents the Ångström exponent.

The accuracy evaluation indexes include R and RMSE, where RMSE is as shown in Equation (6).

$$RMSE = \sqrt{\frac{1}{N}\sum_{i=1}^{N}(\tau(MOD\,AOD)_i - \tau(AERONET)_i)^2} \tag{6}$$

where $\tau(MOD\,AOD)$ and $\tau(AERONET)$ represent the AOD from MOD AOD and $AERONET$, respectively.

4. Results

4.1. LightGBM Training and Processing Results

We constructed and trained the three LightGBM models separately and combined them with 10-fold cross-validation; the sample size, R2, and independent input variables are listed in Table 2. Then, each of the three LightGBM models was used to predict the missing MOD AOD, and we superimposed the prediction results (where the overlap of the pixels is weighted according to R2); the results for 1 January 2018 are listed in Figure 5.

Table 2. LightGBM results and other variables.

Group	Auxiliary Independent Variables	n	R^2
MOD AOD-MYD AOD	TLML, SPEED, ZM, QSH, PBLH, NDVI,	2,112,108	0.964
MOD AOD-MAIAC AOD	POP, RLG, DOY, Slope, Aspect and	4,226,536	0.975
MOD AOD-AHI AOD	Elevation	5,784,070	0.956

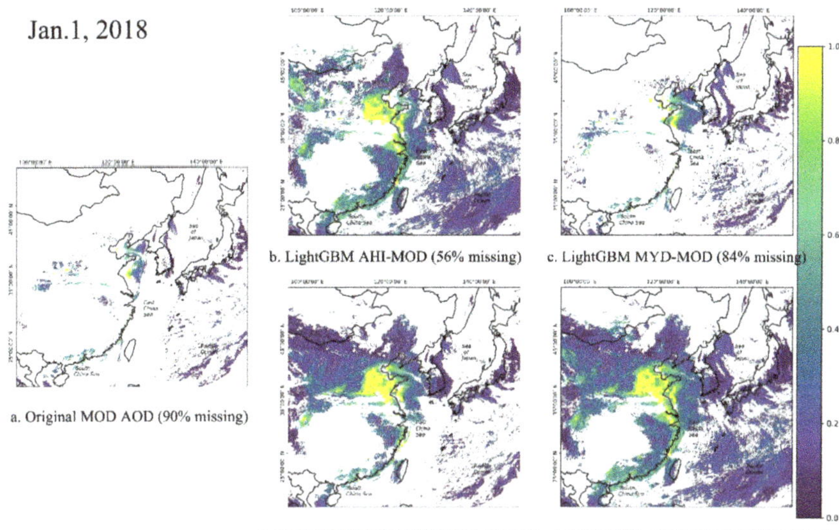

Figure 5. MOD AOD is recovered from multisource AOD data and auxiliary data after fitting by LightGBM (2018.1.1). Here, (**a**) shows the original MOD AOD data (90% missing AOD); (**b**) shows the MOD AOD (56% missing AOD) after AHI AOD, and the auxiliary data were recovered by LightGBM; (**c**) shows the MOD AOD after combining MYD AOD and the auxiliary data after LightGBM recovery (84% missing AOD); (**d**) shows the MOD AOD (66% missing AOD) after combining MAIAC AOD and the auxiliary data after LightGBM recovery; (**e**) shows the result of calculating the weighted average of the overlapping parts of (**b**), (**c**) and (**d**) (47% missing AOD). The legend is the value range of the MOD AOD.

In Table 2, it can be seen that all of the auxiliary variables were involved in the training of the three groups of LightGBM models, and the R^2 of the 10-fold cross-validation fitting effect exceeded 0.95. Additionally, in 1 January 2018, the MOD AOD gap was filled by MYD AOD, MAIAC AOD, and AHI AOD. Among them, AHI AOD contributed the largest quantity of AOD data. The AOD missing rate predicted by AHI AOD decreased from 90% to 56%. After calculating the weighted average of the overlapping parts, the AOD missing rate dropped to 47%.

4.2. Comparison between MOD AOD Recovered by Different Methods and AERONET

We compared the AOD data recovered by different methods with AERONET: 1. The original MOD AOD data and AERONET. 2. The first step of the TWS (LightGBM) was used to calculate the recovered AOD and AERONET comparison. 3. Using spatiotemporal kriging interpolation to interpolate the MOD AOD, we then compared the AOD results with AERONET data. 4. The TWS calculation results were compared with AERONET. To evaluate the effect of the TWS model more carefully, the accuracy of the comparison was divided into all of the AOD data parts (including the recovered part of the AOD and the original MOD AOD part) and a separate AOD recovery part (excluding the original MOD AOD data), as shown in Figures 6 and 7.

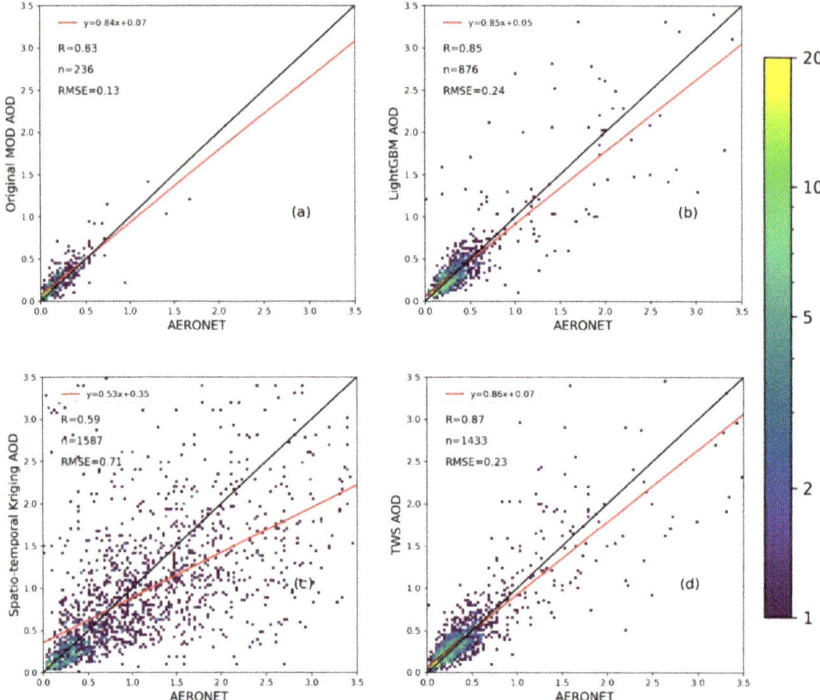

Figure 6. Comparison of MOD AOD recovered by different methods (including the recovered MOD AOD and original MOD AOD) and AERONET. (**a**) Comparison of the original MOD AOD and AERONET. (**b**) Comparison of the MOD AOD recovered by LightGBM and AERONET. (**c**) Comparison of the MOD AOD recovered by spatiotemporal kriging interpolation and AERONET. (**d**) Comparison of the MOD AOD recovered by TWS with AERONET. The solid red line represents the regression line; the solid black line is the 1:1 line. The color bars represent the density of the points.

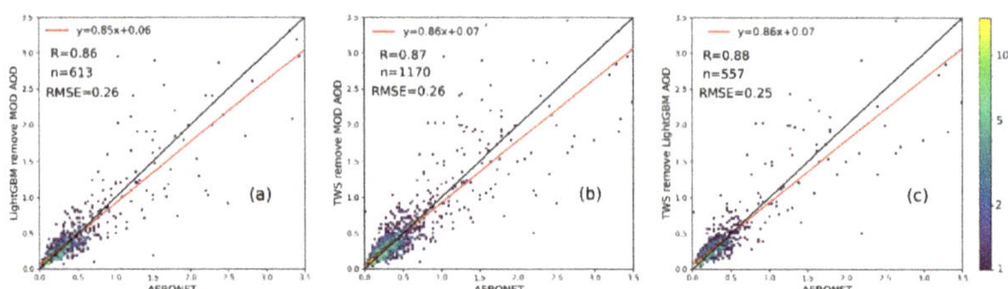

Figure 7. Comparison of MOD AOD and AERONET recovered by different methods. (**a**). Comparison of the recovered MOD AOD of LightGBM (excluding the original MOD AOD part) and AERONET. (**b**). Comparison of the MOD AOD recovered by TWS (excluding the original MOD AOD part) and AERONET. (**c**). Comparison of the MOD AOD recovered by TWS (excluding the original MOD AOD and LightGBM recovered MOD AOD) and AERONET. The solid red line represents the regression line; the solid black line is the 1:1 line. The colored bars represent the density of the points.

As shown in Figures 6 and 7, the number of matching points of MOD AOD and AERONET for reference are 263, the R is 0.83, and the Root Mean Square Error (RMSE) is 0.13. In the comparison of all of the AOD pixel values, LightGBM has the least number of matching points (n = 876), and although the number of matching points in the spatiotemporal kriging interpolation is the largest (1587), the quality according to the R and RMSE (0.59, 0.71) is not as good as that of LightGBM (0.85, 0.24), while TWS (R = 0.87, RMSE = 0.23) maintains value of the R with LightGBM and the reference and the quality of RMSE while also obtaining a larger number of matching points (1433). In the comparison of the AOD recovery part, we computed the results of the TWS recovery after removing MOD AOD (R = 0.87, RMSE = 0.26) and LightGBM (R = 0.88, RMSE = 0.25), and the R and the indicators of RMSE were removed from LightGBM MOD AOD (R = 0.86, RMSE = 0.26), which is consistent; the R is consistent with the reference (the difference in the RMSE index is related to the number and distribution of the reference samples). It can be seen from the results that TWS not only utilizes the information volume of the multisource AOD data, but also absorbs the advantages of AOD spatiotemporal information. In the case of increasing the number of matching points, the R can still maintain a high quality, which indicates that the TWS is reliable.

To further verify the effectiveness of TWS, we regridded the original MOD AOD by 5 × 5 AOD pixels size, and set the existing value in the grid center as a forced-missing AOD. Then, we used 3 × 3 grid TWS to regenerate the forced-missing MOD AOD. A validation between the regenerated MOD AOD and the original effective MOD AOD is shown in Figure 8.

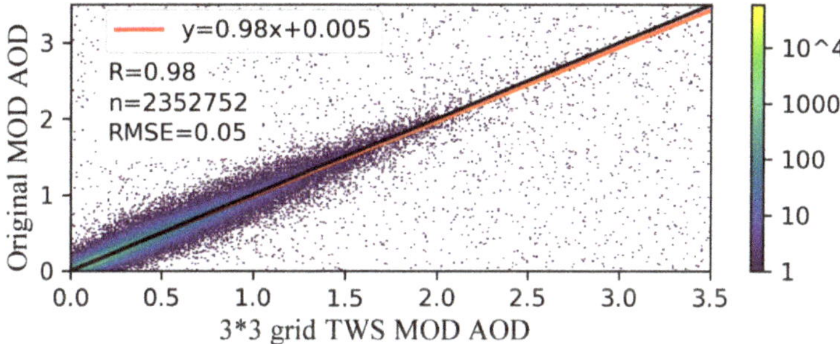

Figure 8. Comparison of the regenerated MOD AOD by 3 × 3 TWS and the original MOD AOD. The solid red line represents the regression line; the solid black line is the 1:1 line. The colored bars represent the density of the points.

As shown in Figure 8, the number of regenerated MOD AOD is 2352752. After restoring the missing AOD by 3 × 3 grid TWS, the validation process results in R = 0.98 and RMSE = 0.05 between the regenerated MOD AOD and the original effective MOD AOD. These results show that the 3 × 3 grid TWS also maintains good stability and accuracy in recovering a large number of missing MOD AOD pixels. This verifies the reliability of the TWS.

4.3. TWS Recovered the Performance with Different Moving Windows

The missing rate for MOD AOD was calculated by the ratio of the MOD AOD pixels and the total number of pixels in the study area, as shown in Figure 9. The MOD AOD missing rate was set to between 0 and 1. The recovery of MOD AOD requires higher accuracy and a lower MOD AOD missing rate to achieve its goal. Although the MOD AOD after the spatiotemporal kriging interpolation processing had no AOD data missing, the accuracy could not reach the application level. Therefore, the comparison of the MOD AOD missing rate was conducted in different windows of the TWS (3 × 3 window, adaptive window and 7 × 7 window). According to the statistics of the original MOD AOD data and the MOD AOD results recovered by TWS, the annual average missing rate of the original MOD AOD exceeded 0.8. After the first step of the TWS LightGBM calculation, the average annual missing rate of MOD AOD decreased from 0.8 to 0.4, and after 3 × 3 restoration of the window, the annual average missing rate of MOD AOD decreased from 0.4 to 0.1; additionally, the result calculated after the 7 × 7 window (0.06) showed the smallest annual average missing rate of MOD AOD.

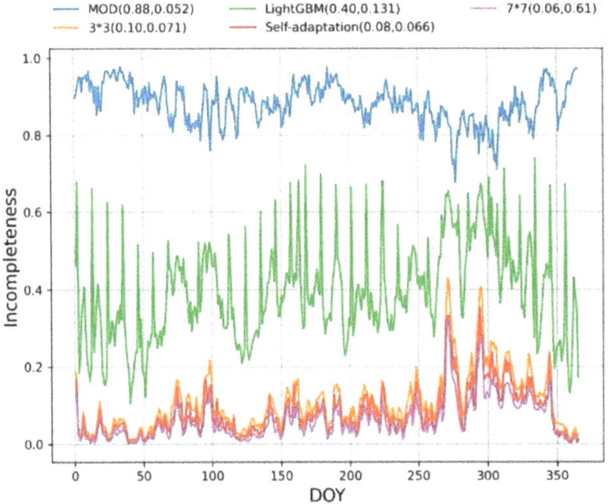

Figure 9. Time series plot of daily AOD coverage over study areas in 2018 for MOD, LightGBM, 3 × 3, self-adaption and 7 × 7. MOD represents the original MOD AOD, LightGBM represents LightGBM recovered MOD AOD, 3 × 3 represents the 3 × 3 grid moving window TWS recovered MOD AOD, self-adaption represents the self-adaption moving window TWS recovered MOD AOD, and 7 × 7 represents the 7 × 7 moving window TWS recovered MOD AOD. The numbers in parentheses represent the average and standard deviation of the empty AOD coverage.

Furthermore, in 2018, the standard deviation of the missing rate of MOD AOD after LightGBM alone was 0.131. However, the standard deviation of the MOD AOD missing rate of the TWS treatment was smaller than 0.08, which shows that LightGBM alone relies on only multisource AOD data. After processing by LightGBM alone, there is still a large quantity of missing AOD data. In contrast, a complete TWS combined with spatial and spatiotemporal information can reduce the missing rate of MOD AOD.

According to Table 3 and Figure 10, the missing rate of MOD AOD, R, and the calculation efficiency all change with changes in the size of the moving window. Among them, the 7 × 7 grid has the lowest R and the largest RMSE, 0.78 and 0.32, respectively. The adaptive R and RMSE are 0.79 and 0.3, respectively. The 7 × 7 grid and adaptive R decrease compared to the 3 × 3 window, while the RMSE increases. The adaptive network's calculation time of the grid is the largest, i.e., 4.2 times that of the 3 × 3 grid, while the 7 × 7 grid is 2.7 times that of the 3 × 3 grid. The above data show that with the expansion in the window size, the result R from the recovery of the MOD AOD decreases, while the RMSE increases. A possible reason for this is that the spatial and temporal variability of the MOD AOD increases with the size of the moving window. Moreover, the change in the size of the moving window also significantly affects the amount of calculation.

Table 3. Performance comparison of 3 different moving windows.

Windows	R (Total)	Incompleteness (%)	Time Ratio (%)
3 × 3 grid	0.85	10	100
7 × 7 grid	0.78	6	225
Self-adaption grid	0.79	8	423

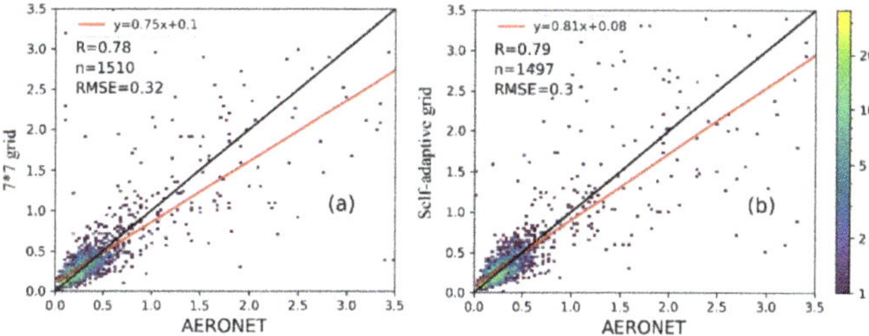

Figure 10. Comparison of TWS recovered MOD AOD (including the recovered MOD AOD and original MOD AOD) and AERONET in different moving windows. **a**. Comparison of the 7 × 7 moving window TWS recovery MOD AOD and AERONET. **b** Comparison of the self-adaption moving window TWS recovery MOD AOD and AERONET. The solid red line represents the regression line; the solid black line is the 1:1 line. The colored bars represent the density of points.

4.4. Analysis of the Spatiotemporal Characteristics of MOD AOD Recovered by TWS

Combining the recovery results of the MOD AOD in the 3 × 3 window of the TWS and the spatiotemporal kriging interpolation results of the MOD AOD, the annual average results of the MOD AOD after recovery were calculated and compared with the annual average results of the original MOD AOD (Figure 11). The following can be found in Figure 11: (1). There were still some gaps in the annual average map of the original MOD AOD (the position of the red circle 1). Compared with Figure 1 (land use), the red circle is mainly brighter, bare land, which confirmed that the DT algorithm and the DB algorithm had poor AOD data inversion in relatively bright areas. The annual average result of the MOD AOD recovery in the 3 × 3 window of TWS and the annual average result of the MOD AOD spatiotemporal Kriging interpolation filled the gaps of the AOD data in the red circle 1. (2). The maximum value of the original annual average result of the MOD AOD is too large in Figure 11 (the maximum AOD value was 3). (3). The maximum value in the annual average result of MOD AOD in the 3 × 3 window of TWS decreased to 0.64 and the annual average result of the spatiotemporal kriging interpolation of MOD AOD decreased to 0.82. (4). The average value in the annual average results of the original MOD AOD, spatiotemporal kriging interpolation and TWS were 0.23, 0.34 and 0.27 respectively. The original MOD AOD data had a large number of missing AOD pixels (the missing rate in Figure 11a was 2%). There was a lack of sufficient AOD pixels to average the minimum and maximum values in the original MOD AOD, which ultimately led to the maximum value in the original MOD AOD annual average result being too large (the maximum AOD value was 3), and the average value in the original MOD AOD annual average result was low (the average AOD value was 0.23). (5). Comparing red circle 2, the annual average results of the original MOD AOD and the spatiotemporal Kriging interpolation of the MOD AOD are higher. The annual average results of the restoration of MOD AOD in the 3 × 3 window of TWS retained the original MOD AOD spatial characteristics of the annual average results and reduced the annual average of MOD AOD. Moreover, in the Pacific region, the annual average results of the restoration of the TWS 3 × 3 window MOD AOD were higher than the original annual average results of the original MOD AOD. In the original annual average results of the MOD AOD, the reason why the AOD data gap in red circle 1 was filled is that the 3 × 3 window of the TWS and the spatiotemporal kriging interpolation method filled the AOD data gap to a large extent. The reason for this was that the MOD AOD gap was filled, and the MOD AOD annual average result was more fully calculated. The maximum value of the original MOD AOD annual average result was reduced. In addition, due to the lack of accurate prediction of local features by the spatiotemporal kriging interpolation algorithm, the annual average result of the MOD AOD spatiotemporal kriging interpolation was higher than the average annual result of the restoration of the TWS 3 × 3 window MOD AOD.

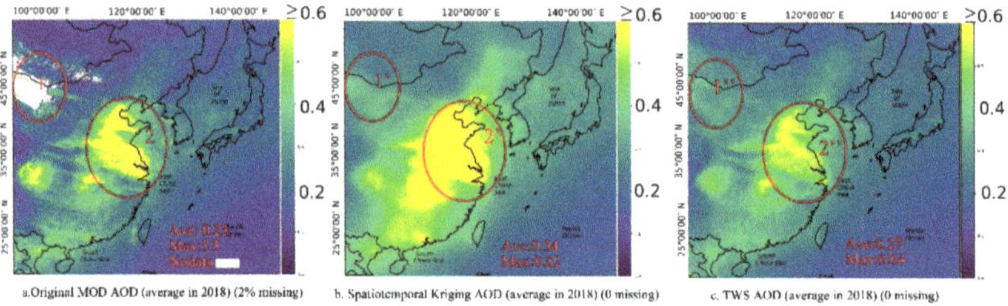

Figure 11. The average annual MOD AOD distribution in 2018. (**a**). Annual average of the original MOD AOD (2% missing). (**b**). The MOD AOD average of the spatiotemporal kriging interpolation recovery (missing 0). (**c**). The 3 × 3 moving window TWS recovered the MOD AOD annual average (missing 0). The red fonts Ave and Max represent the average and maximum values of the AOD annual average graph, respectively. The white part represents nodata. The color bar represents the MOD AOD value.

We compared the results of the TWS 3 × 3 window MOD AOD recovery with the original MOD AOD data by a monthly average (Figure 12). In Figure 12, we marked the missing rate, average and maximum of the monthly average of the original MOD AOD and the monthly average of TWS AOD for each month. The monthly average maximum value of TWS AOD was smaller than the original monthly average maximum value of MOD AOD. The average range of the monthly average results of TWS AOD (0.17–0.24) was also smaller than the average monthly average results of the original MOD AOD (0.18–0.36). In addition, the TWS AOD monthly average result also accurately retained the high value area of the original MOD AOD monthly average result (in the yellow box).

On this basis, in the yellow box area in Figure 12 (112.7° E–125.2° E, 32.5° N–42.1° N), we calculated the monthly average and maximum AOD values of the original MOD AOD and TWS AOD in this area, as well as the monthly average AERONET AOD at the same place (Figure 13). In the yellow box area, the maximum of the monthly average original MOD AOD result was greater than 2. The maximum of the monthly average TWS AOD result was lower than the maximum of the monthly average original MOD AOD result. Moreover, the largest average value of the monthly average TWS AOD results was in June. Specifically, there was an upward trend from January to June and a downward trend from June to December. In addition, in the yellow box, there are seven AERONET ground stations. We calculated the monthly average of these stations. The monthly average trend of MOD AOD after TWS recovery was also consistent with the monthly average AERONET AOD trend. A similar trend was shown by Song et al. [56] for the North China Plain in 2018.

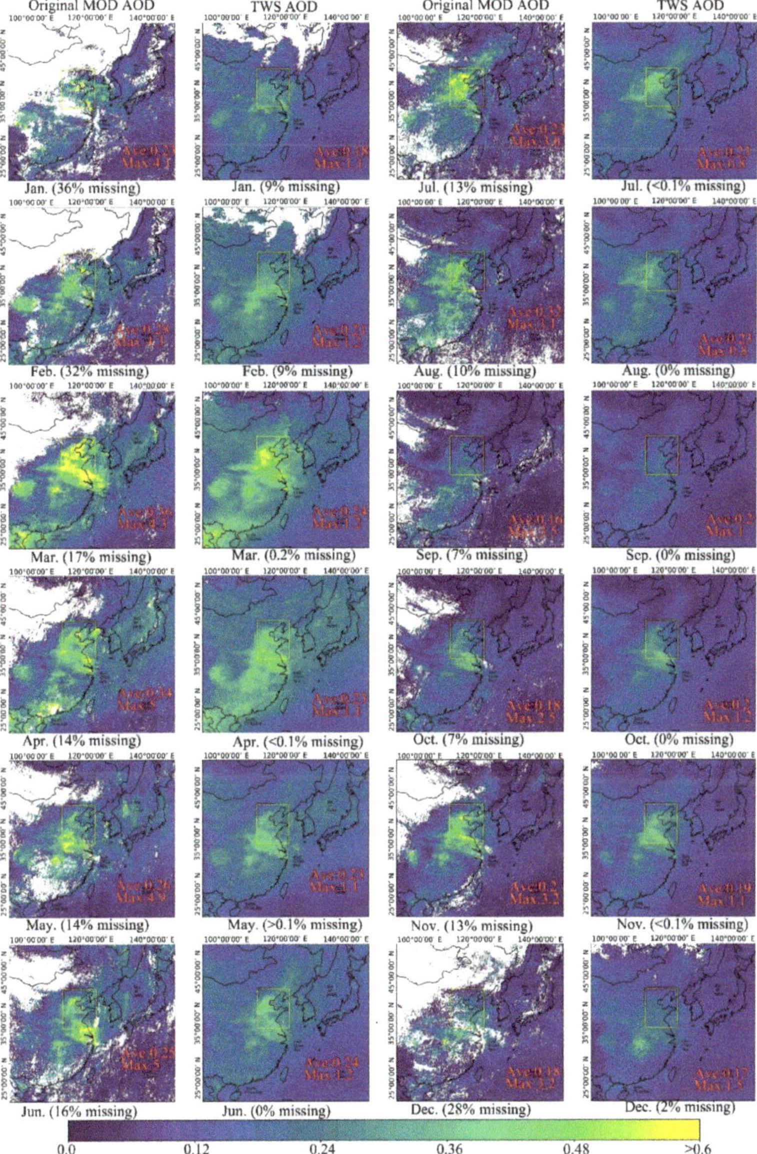

Figure 12. Monthly average of the original MOD AOD and 3 × 3 moving window TWS recovered MOD AOD (each month includes the missing rate of MOD AOD). The red fonts Ave and Max represent the average and maximum values of the AOD monthly average graph, respectively. The white part represents no data. The yellow box area represents the sampling area in Figure 13. The color bar represents the MOD AOD value.

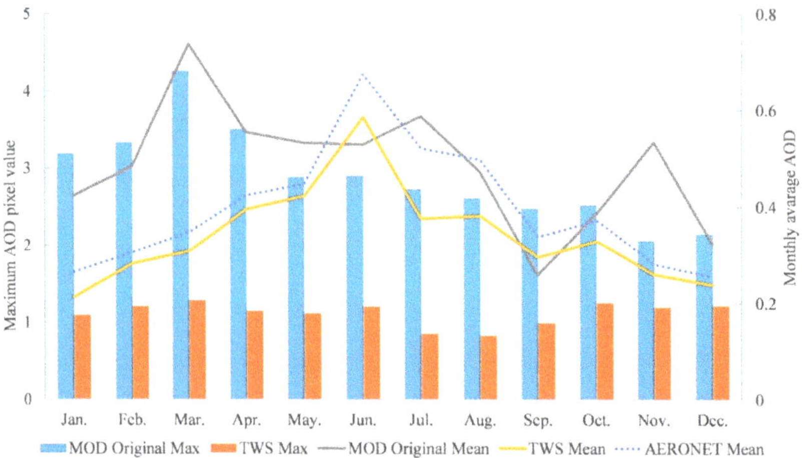

Figure 13. The broken line represents the monthly average original MOD AOD and the monthly average MOD AOD restored by TWS, respectively, and the dotted line represents the monthly average AERONET AOD, with the ordinate on the left. The histogram represents the maximum value (monthly) of the original MOD AOD and the MOD AOD recovered by TWS, respectively, with the ordinate on the right. The above data is from the yellow box area of Figure 12.

5. Discussion

5.1. Comparison of TWS and Other MOD AOD Recovery Models

The recovery of missing satellite AOD product data is of great significance to atmospheric pollution research. Recently, many methods have been used to study the recovery of missing data from satellite AOD products. This study selected the same approach to recover missing MOD AOD data and made a comparison in Table 4. The results of the various methods in Table 4 were compared with AERONET. Based on this comparison, the improvements in R compared to the MOD AOD and AERONET recovered by the proposed method and the R compared to the original MOD AOD and AERONET were not obvious (the R of the MOD AOD and AERONET recovered by the TWS recovery was the highest). In the comparison of the missing rate of MOD AOD, the missing rate of MOD AOD recovered by TWS was the lowest (0.1). Additionally, in the different methods in Table 4, the missing rate of the MOD AOD recovered by TWS had the largest decreased missing rate difference compared to the original MOD AOD (0.78). The improved difference (R) of the 3 × 3 window TWS method was not significantly different from other methods. However, the decreased missing rate difference (%) of the 3 × 3 window TWS method was significantly different from other methods. The main reasons are as follows: (1.) The 3 × 3 window TWS introduced multisource datasets (MYD AOD, MAIAC AOD, AHI AOD). With TWS, the first step is to use the spatial complement of AOD data sets with different algorithms and data collection times. The AOD missing rate dropped from 88% to 40%. In Figure 9, the decreased missing rate difference (%) is 48%. (2.) The second step of the 3 × 3 window TWS is to make reasonable use of the spatiotemporal relationship of AOD, under the optimization of moving window and buffer factor. The AOD missing rate decreased from 40% to 10% (the decreased missing rate difference (%) was 30%). Although the direct comparison of the decreased missing rate difference had certain limitations, it also showed stability and excellent performance for the TWS.

Table 4. Comparison of the MOD AOD data recovery methods.

Method	Original Missing Rate (%)	Improved Missing Rate (%)	Decreased Missing Rate Difference (%)	Original R	Improved R	Improved Difference (R)	Source
ST-AVM	80	60	20	0.89	0.87	−0.02	[34]
NWRL	~70	~60	~10	0.77	0.78	+0.01	[33]
*	89	75	14	0.93	0.91	−0.02	[28]
TWS (3 × 3)	88	10	78	0.83	0.87	+0.04	Our paper

Note: ~ indicates a lack of accurate data in the cited article. * indicates a lack of method name in the cited article.

5.2. TWS Recovery MOD AOD Performance Discussion

The MOD AOD after TWS processing can obtain a higher improved R and lower AOD missing rate because it takes full advantage of the rich data volume of multisource data and the high local spatiotemporal autocorrelation of the AOD itself. A large amount of research has confirmed that multisource data can easily introduce data noise. However, based on the data statistics, we chose LightGBM to build a MOD AOD prediction model, which can make full use of the characteristics of different AOD data and reduce the data noise. From the comparison between LightGBM and AERONET, it can be seen that the LightGBM model does not introduce much data noise (all R = 0.85, R = 0.86 after removing MOD AOD).

Moreover, we developed MOD AOD recovery measures based on moving small windows by combining MOD AOD spatial data and spatiotemporal data when generating the statistics. The setting of the small window is used to ensure a high correlation of AOD in the small window. MOD AOD recovery measures set three MOD AOD recovery modes, and use the adaptive space and spatiotemporal buffering methods. Different calculation modes were set based on the temporal and spatial distribution of valid AOD information, to enable the calculation to be more reliable when recovering the AOD value. In this way, it can avoid the introduction of excessive data noise. The index was used to determine the local area of the autocorrelation, and the mathematical expectation and R were introduced to slow down the spatiotemporal difference; then, we determined the spatial and spatiotemporal buffer. Spatial and spatiotemporal buffering can more accurately improve the R of the moving small windows to recover the MOD AOD missing data. These settings all ensure the accuracy of the MOD AOD recovery and reduce the data loss rate of MOD AOD (R = 0.87 compared to MOD AOD and AERONET in the 3 × 3 window. The average daily loss rate of MOD AOD was 10%, whereas the adaptive window of the MOD AOD and AERONET comparison was R = 0.79, the average daily missing rate of MOD AOD was 8%, the window of the 7 × 7 window MOD AOD and AERONET comparison was R = 0.78, and the average daily missing rate of MOD AOD was 6%). In different applications, different window sizes can be chosen to meet different needs because the moving window size is variable. For example, to obtain a lower MOD AOD data loss rate, a larger moving window in TWS can be selected. The 7 × 7 window in the 2018 experiment can limit the average daily loss rate of MOD AOD to 6%. Therefore, moving the window size can adjust this advantage and make the TWS method more flexible. Moreover, if the missing MOD AOD data rate is not 0, the iterative approach to the TWS method can be used, which gradually reduces the missing MOD AOD data rate to 0. Of course, it is also possible to use spatial interpolation based on the results of MOD AOD processed by TWS to reduce the missing rate of MOD AOD to 0. Because TWS is based on sufficient data statistics on AOD data and uses AOD spatial autocorrelation, the TWS method can, in general, be applied to the missing data of AHI AOD, MAIAC AOD, MYD AOD and other remote sensing products with spatial correlation and time correlation. Finally, the MOD AOD recovered by TWS cannot be studied and used on a global scale because the AHI sensor is carried on a geosynchronous orbit satellite.

5.3. TWS Recovery MOD AOD

In the results of MOD AOD recovery in the study area in 2018 (Figure 11), we found that the areas with higher AOD were mainly concentrated in North China, the Central China Plain, the Yangtze River Delta and the Sichuan Basin; in northern Vietnam, the Japanese Islands and the Korean Peninsula, the AOD was lower (see Figure 12). Because of the dry weather conditions of the Red River Delta, in addition to local traffic and industrial pollution, there was a relatively obvious pollutant transmission process, and higher AOD distributions existed in southern China and northern Vietnam in March and April [57]. Moreover, there was also a higher monthly average value of AOD near the North China Plain and Shandong Peninsula in China which spread to the sea. Furthermore, in November, December and January, the pollutant diffusion capacity of North China, the Central China Plain and the Yangtze River Delta was more obvious during the influence of the winter monsoon [58,59]. Eventually, the mean monthly AOD of North China, the Central China Plain, the Yangtze River Delta and the East China Sea increased. Overall, the high AOD area did not cover the Korean Peninsula or the Japanese Islands. Although some of the pollutants might have reached the Korean Peninsula and the Japanese Archipelago region through the atmospheric transmission process, most of the pollutant transmission still stopped in the offshore area of China.

6. Conclusions

A high-precision, low AOD missing rate MOD AOD recovery result is of great help in measuring the spatial distribution of air pollutants, continuous monitoring, climate change and other related research. In this paper, the TWS model was constructed by multisource AOD data, LightGBM, spatial interpolation and STW, which were used for the large-scale recovery of data missing from MOD AOD. The results show that the TWS model can guarantee the accuracy of the recovered MOD AOD ($R = 0.87$). Moreover, compared with other methods, TWS greatly reduces the missing rate of the MOD AOD data (the missing rate of MOD AOD in the 3×3 window dropped from the original 88% to 10%). Moreover, after the missing information is added, the changes in the local AOD start to show more obvious high and low value details, for example, the AOD average, maximum and minimum of the original MOD AOD missing area in the AOD annual average map. TWS proves the spatial complementarity of multisource AOD data and the spatiotemporal relationship of the AOD data, which is very important when recovering the AOD data. In follow-up research, we will use other data sets to expand the applicability of the TWS method, for example, using GOES-16 ABI AOD data to restore AOD on the American continent. Moreover, we will use deep learning to recover areas in which the loss of AOD spatiotemporal information is severe, for example, in scenario 3 (Pass) in Figure 2, the moving window has missing AOD information for three consecutive days.

Author Contributions: Methodology, Y.C.; validation, Y.C.; formal analysis, Y.C.; writing—original draft preparation, Y.C.; writing—review and editing, Z.W., Y.R. and K.L. All authors have read and agreed to the published version of the manuscript.

Funding: This work was supported by the National Science Foundation of China (31670645, 31972951, 41801182, and 41807502), the National Social Science Fund (17ZDA058), the National Key Research Program of China (2016YFC0502704), Fujian Provincial Department of S&T Project (2018T3018), the Strategic Priority Research Program of the Chinese Academy of Sciences (XDA23020502), the Ningbo Municipal Department of Science and Technology (2019C10056), the Xiamen Municipal Department of Science and Technology (3502Z20130037 and 3502Z20142016), the Key Laboratory of Urban Environment and Health of CAS (KLUEH-C-201701), the Key Program of the Chinese Academy of Sciences (KFZDSW-324), and the Youth Innovation Promotion Association CAS (2014267). Open-end Fund (2020KX03).

Acknowledgments: We are grateful to the anonymous reviewers for their constructive suggestions. The authors would like to express our gratitude to the National Aeronautics and Space Administration (NASA) for providing the MODIS AOD, MAIAC, MERRA2, SRTM and NDVI products, gratitude to the Japan Meteorological Agency (JMA) and Center of Environmental Remote Sensing, Chiba University (CEReS) for providing the AHI AOD product, and the Principle Investigators for establishing and maintaining the AERONET sites.

Conflicts of Interest: The authors declare no conflict of interest

References

1. Hallquist, M.; Wenger, J.C.; Baltensperger, U.; Rudich, Y.; Simpson, D.; Claeys, M.; Dommen, J.; Donahue, N.M.; George, C.; Goldstein, A.H.; et al. The formation, properties and impact of secondary organic aerosol: Current and emerging issues. *Atmos. Chem. Phys.* **2009**, *9*, 5155–5236. [CrossRef]

2. Mahowald, N. Aerosol Indirect Effect on Biogeochemical Cycles and Climate. *Science* **2011**, *334*, 794–796. [CrossRef]
3. Dubovik, O.; Holben, B.; Eck, T.F.; Smirnov, A.; Kaufman, Y.J.; King, M.D.; Tanré, D.; Slutsker, I. Variability of Absorption and Optical Properties of Key Aerosol Types Observed in Worldwide Locations. *J. Atmos. Sci.* **2002**, *59*, 590–608. [CrossRef]
4. Ramanathan, V.; Crutzen, P.J.; Kiehl, J.T.; Rosenfeld, D. Aerosols, climate, and the hydrological cycle. *Science* **2001**, *294*, 2119–2124. [CrossRef] [PubMed]
5. Adhikary, B.; Kulkarni, S.; Dallura, A.; Tang, Y.; Chai, T.; Leung, L.R.; Qian, Y.; Chung, C.E.; Ramanathan, V.; Carmichael, G.R. A regional scale chemical transport modeling of Asian aerosols with data assimilation of AOD observations using optimal interpolation technique. *Atmos. Environ.* **2008**, *42*, 8600–8615. [CrossRef]
6. Xin, J.; Wang, Y.; Li, Z.; Wang, P.; Hao, W.; Nordgren, B.; Wang, S.; Liu, G.; Wang, L.; Wen, T.; et al. Aerosol optical depth (AOD) and Ångström exponent of aerosols observed by the Chinese Sun Hazemeter Network from August 2004 to September 2005. *J. Geophys. Res.* **2007**, *112*. [CrossRef]
7. Holben, B.N.; Eck, T.F.; Slutsker, I.; Tanré, D.; Buis, J.P.; Setzer, A.; Vermote, E.; Reagan, J.A.; Kaufman, Y.J.; Nakajima, T.; et al. AERONET—A Federated Instrument Network and Data Archive for Aerosol Characterization. *Remote Sens. Environ.* **1998**, *66*, 1–16. [CrossRef]
8. Giles, D.M.; Sinyuk, A.; Sorokin, M.G.; Schafer, J.S.; Smirnov, A.; Slutsker, I.; Eck, T.F.; Holben, B.N.; Lewis, J.R.; Campbell, J.R.; et al. Advancements in the Aerosol Robotic Network (AERONET) Version 3 database—Automated near-real-time quality control algorithm with improved cloud screening for Sun photometer aerosol optical depth (AOD) measurements. *Atmos. Meas. Tech.* **2019**, *12*, 169–209. [CrossRef]
9. Zhao, C.; Liu, Z.; Wang, Q.; Ban, J.; Chen, N.X.; Li, T. High-resolution daily AOD estimated to full coverage using the random forest model approach in the Beijing-Tianjin-Hebei region. *Atmos. Environ.* **2019**, *203*, 70–78. [CrossRef]
10. Remer, L.A.; Kaufman, Y.J.; Tanré, D.; Mattoo, S.; Chu, D.A.; Martins, J.V.; Li, R.R.; Ichoku, C.; Levy, R.C.; Kleidman, R.G.; et al. The MODIS Aerosol Algorithm, Products, and Validation. *J. Atmos. Sci.* **2005**, *62*, 947–973. [CrossRef]
11. Levy, R.C.; Mattoo, S.; Munchak, L.A.; Remer, L.A.; Sayer, A.M.; Patadia, F.; Hsu, N.C. The Collection 6 MODIS aerosol products over land and ocean. *Atmos. Meas. Tech.* **2013**, *6*, 2989–3034. [CrossRef]
12. Hsu, N.C.; Jeong, M.J.; Bettenhausen, C.; Sayer, A.M.; Hansell, R.; Seftor, C.S.; Huang, J.; Tsay, S.C. Enhanced Deep Blue aerosol retrieval algorithm: The second generation. *J. Geophys. Res. Atmos.* **2013**, *118*, 9296–9315. [CrossRef]
13. Levy, R.C.; Remer, L.A.; Kleidman, R.G.; Mattoo, S.; Ichoku, C.; Kahn, R.; Eck, T.F. Global evaluation of the Collection 5 MODIS dark-target aerosol products over land. *Atmos. Chem. Phys.* **2010**, *10*, 10399–10420. [CrossRef]
14. Hyer, E.J.; Reid, J.S.; Zhang, J. An over-land aerosol optical depth data set for data assimilation by filtering, correction, and aggregation of MODIS Collection 5 optical depth retrievals. *Atmos. Meas. Tech.* **2011**, *4*, 379–408. [CrossRef]
15. Lyapustin, A.; Wang, Y.; Korkin, S.; Huang, D. MODIS Collection 6 MAIAC algorithm. *Atmos. Meas. Tech.* **2018**, *11*, 5741–5765. [CrossRef]
16. Lyapustin, A.; Wang, Y.; Laszlo, I.; Kahn, R.; Korkin, S.; Remer, L.; Levy, R.; Reid, J.S. Multiangle implementation of atmospheric correction (MAIAC): 2. Aerosol algorithm. *J. Geophys. Res. Atmos.* **2011**, *116*. [CrossRef]
17. Wei, J.; Huang, W.; Li, Z.; Xue, W.; Peng, Y.; Sun, L.; Cribb, M. Estimating 1-km-resolution PM2.5 concentrations across China using the space-time random forest approach. *Remote Sens. Environ.* **2019**, *231*, 111221. [CrossRef]
18. Kikuchi, M.; Murakami, H.; Suzuki, K.; Nagao, T.M.; Higurashi, A. Improved Hourly Estimates of Aerosol Optical Thickness Using Spatiotemporal Variability Derived From Himawari-8 Geostationary Satellite. *IEEE Trans. Geosci. Remote Sens.* **2018**, *56*, 3442–3455. [CrossRef]

19. Letu, H.; Nagao, T.M.; Nakajima, T.Y.; Riedi, J.; Ishimoto, H.; Baran, A.J.; Shang, H.; Sekiguchi, M.; Kikuchi, M. Ice Cloud Properties From Himawari-8/AHI Next-Generation Geostationary Satellite: Capability of the AHI to Monitor the DC Cloud Generation Process. *IEEE Trans. Geosci. Remote Sens.* **2019**, *57*, 3229–3239. [CrossRef]
20. Zhan, Y.; Luo, Y.; Deng, X.; Chen, H.; Grieneisen, M.L.; Shen, X.; Zhu, L.; Zhang, M. Spatiotemporal prediction of continuous daily PM2.5 concentrations across China using a spatially explicit machine learning algorithm. *Atmos. Environ.* **2017**, *155*, 129–139. [CrossRef]
21. Wei, J.; Li, Z.; Cribb, M.; Huang, W.; Xue, W.; Sun, L.; Guo, J.; Peng, Y.; Li, J.; Lyapustin, A.; et al. Improved 1 km resolution $PM_{2.5}$ estimates across China using enhanced space–time extremely randomized trees. *Atmos. Chem. Phys.* **2020**, *20*, 3273–3289. [CrossRef]
22. Ghotbi, S.; Sotoudeheian, S.; Arhami, M. Estimating urban ground-level PM10 using MODIS 3km AOD product and meteorological parameters from WRF model. *Atmos. Environ.* **2016**, *141*, 333–346. [CrossRef]
23. Mhawish, A.; Banerjee, T.; Broday, D.M.; Misra, A.; Tripathi, S.N. Evaluation of MODIS Collection 6 aerosol retrieval algorithms over Indo-Gangetic Plain: Implications of aerosols types and mass loading. *Remote Sens. Environ.* **2017**, *201*, 297–313. [CrossRef]
24. Zhang, R.; Di, B.; Luo, Y.; Deng, X.; Grieneisen, M.L.; Wang, Z.; Yao, G.; Zhan, Y. A nonparametric approach to filling gaps in satellite-retrieved aerosol optical depth for estimating ambient PM2.5 levels. *Environ. Pollut.* **2018**, *243*, 998–1007. [CrossRef] [PubMed]
25. Lv, B.; Hu, Y.; Chang, H.H.; Russell, A.G.; Bai, Y. Improving the Accuracy of Daily PM2.5 Distributions Derived from the Fusion of Ground-Level Measurements with Aerosol Optical Depth Observations, a Case Study in North China. *Environ. Sci. Technol.* **2016**, *50*, 4752–4759. [CrossRef] [PubMed]
26. Chi, Y.; Zuo, S.; Ren, Y.; Chen, K. The Spatiotemporal Pattern of the Aerosol Optical Depth (AOD) on the Canopies of Various Forest Types in the Exurban National Park: A Case in Ningbo City, Eastern China. *Adv. Meteorol.* **2019**, *2019*, 12. [CrossRef]
27. Song, Z.; Fu, D.; Zhang, X.; Han, X.; Song, J.; Zhang, J.; Wang, J.; Xia, X. MODIS AOD sampling rate and its effect on PM2.5 estimation in North China. *Atmos. Environ.* **2019**, *209*, 14–22. [CrossRef]
28. Cheng, L.; Li, L.; Chen, L.; Hu, S.; Yuan, L.; Liu, Y.; Cui, Y.; Zhang, T. Spatiotemporal Variability and Influencing Factors of Aerosol Optical Depth over the Pan Yangtze River Delta during the 2014–2017 Period. *Int. J. Environ. Res. Public Health* **2019**, *16*. [CrossRef]
29. Alam, K.; Qureshi, S.; Blaschke, T. Monitoring spatio-temporal aerosol patterns over Pakistan based on MODIS, TOMS and MISR satellite data and a HYSPLIT model. *Atmos. Environ.* **2011**, *45*, 4641–4651. [CrossRef]
30. Van Donkelaar, A.; Martin, R.V.; Levy, R.C.; da Silva, A.M.; Krzyzanowski, M.; Chubarova, N.E.; Semutnikova, E.; Cohen, A.J. Satellite-based estimates of ground-level fine particulate matter during extreme events: A case study of the Moscow fires in 2010. *Atmos. Environ.* **2011**, *45*, 6225–6232. [CrossRef]
31. Huang, C.; Goward, S.N.; Masek, J.G.; Thomas, N.; Zhu, Z.; Vogelmann, J.E. An automated approach for reconstructing recent forest disturbance history using dense Landsat time series stacks. *Remote Sens. Environ.* **2010**, *114*, 183–198. [CrossRef]
32. Pang, J.; Wang, X.; Shao, M.; Chen, W.; Chang, M. Aerosol optical depth assimilation for a modal aerosol model: Implementation and application in AOD forecasts over East Asia. *Sci. Total Environ.* **2020**, *719*, 137430. [CrossRef] [PubMed]
33. Zhang, T.; Zeng, C.; Gong, W.; Wang, L.; Sun, K.; Shen, H.; Zhu, Z.; Zhu, Z. Improving Spatial Coverage for Aqua MODIS AOD using NDVI-Based Multi-Temporal Regression Analysis. *Remote Sens.* **2017**, *9*. [CrossRef]
34. Wang, Y.; Yuan, Q.; Li, T.; Shen, H.; Zheng, L.; Zhang, L. Large-scale MODIS AOD products recovery: Spatial-temporal hybrid fusion considering aerosol variation mitigation. *ISPRS J. Photogramm.* **2019**, *157*, 1–12. [CrossRef]
35. Singh, M.K.; Venkatachalam, P.; Gautam, R. Geostatistical Methods for Filling Gaps in Level-3 Monthly-Mean Aerosol Optical Depth Data from Multi-Angle Imaging SpectroRadiometer. *Aerosol Air Qual. Res.* **2017**, *17*, 1963–1974. [CrossRef]
36. Yang, J.; Hu, M. Filling the missing data gaps of daily MODIS AOD using spatiotemporal interpolation. *Sci. Total Environ.* **2018**, *633*, 677–683. [CrossRef] [PubMed]

37. Tang, Q.; Bo, Y.; Zhu, Y. Spatiotemporal fusion of multiple-satellite aerosol optical depth (AOD) products using Bayesian maximum entropy method. *J. Geophys. Res. Atmos.* **2016**, *121*, 4034–4048. [CrossRef]
38. Nam, J.; Kim, S.-W.; Park, R.J.; Park, J.-S.; Park, S.S. Changes in column aerosol optical depth and ground-level particulate matter concentration over East Asia. *Air Qual. Atmos. Health* **2018**, *11*, 49–60. [CrossRef]
39. Xiao, Q.; Zhang, H.; Choi, M.; Li, S.; Kondragunta, S.; Kim, J.; Holben, B.; Levy, R.C.; Liu, Y. Evaluation of VIIRS, GOCI, and MODIS Collection 6 AOD retrievals against ground sunphotometer observations over East Asia. *Atmos. Chem. Phys.* **2016**, *16*, 1255–1269. [CrossRef]
40. Hsu, N.C.; Si-Chee, T.; King, M.D.; Herman, J.R. Aerosol properties over bright-reflecting source regions. *IEEE Trans. Geosci. Remote.* **2004**, *42*, 557–569. [CrossRef]
41. Xia, X.; Min, J.; Wang, Y.; Shen, F.; Yang, C.; Sun, Z. Assimilating Himawari-8 AHI aerosol observations with a rapid-update data assimilation system. *Atmos. Environ.* **2019**, *215*, 116866. [CrossRef]
42. Jiang, J.H.; Su, H.; Zhai, C.; Wu, L.; Minschwaner, K.; Molod, A.M.; Tompkins, A.M. An assessment of upper troposphere and lower stratosphere water vapor in MERRA, MERRA2, and ECMWF reanalyses using Aura MLS observations. *J. Geophys. Res. Atmos.* **2015**, *120*, 468–485. [CrossRef]
43. Tobin, K.W.; Bhaduri, B.L.; Bright, E.A.; Cheriyadat, A.; Karnowski, T.P.; Palathingal, P.J.; Potok, T.E.; Price, J.R. Automated Feature Generation in Large-Scale Geospatial Libraries for Content-Based Indexing. *Photogramm. Eng. Remote Sens.* **2006**, *72*, 531–540. [CrossRef]
44. Fensholt, R.; Rasmussen, K.; Nielsen, T.T.; Mbow, C. Evaluation of earth observation based long term vegetation trends—Intercomparing NDVI time series trend analysis consistency of Sahel from AVHRR GIMMS, Terra MODIS and SPOT VGT data. *Remote Sens. Environ.* **2009**, *113*, 1886–1898. [CrossRef]
45. Wei, J.; Li, Z.; Sun, L.; Peng, Y.; Zhang, Z.; Li, Z.; Su, T.; Feng, L.; Cai, Z.; Wu, H. Evaluation and uncertainty estimate of next-generation geostationary meteorological Himawari-8/AHI aerosol products. *Sci. Total Environ.* **2019**, *692*, 879–891. [CrossRef]
46. Kaufman, Y.J.; Tanré, D.; Boucher, O. A satellite view of aerosols in the climate system. *Nature* **2002**, *419*, 215–223. [CrossRef]
47. Miller, R.L.; Tegen, I. Climate Response to Soil Dust Aerosols. *J. Climate* **1998**, *11*, 3247–3267. [CrossRef]
48. Zhang, J.; Mucs, D.; Norinder, U.; Svensson, F. LightGBM: An Effective and Scalable Algorithm for Prediction of Chemical Toxicity–Application to the Tox21 and Mutagenicity Data Sets. *J. Chem. Inf. Model* **2019**, *59*, 4150–4158. [CrossRef]
49. Ke, G.; Meng, Q.; Finley, T.; Wang, T.; Chen, W.; Ma, W.; Ye, Q.; Liu, T.-Y. LightGBM: A highly efficient gradient boosting decision tree. In Proceedings of the 31st International Conference on Neural Information Processing Systems, Long Beach, CA, USA, 4–9 December 2017; pp. 3149–3157.
50. Lu, G.Y.; Wong, D.W. An adaptive inverse-distance weighting spatial interpolation technique. *Comput. Geosci. UK* **2008**, *34*, 1044–1055. [CrossRef]
51. Requia, W.J.; Dalumpines, R.; Adams, M.D.; Arain, A.; Ferguson, M.; Koutrakis, P. Modeling spatial patterns of link-based PM2.5 emissions and subsequent human exposure in a large canadian metropolitan area. *Atmos. Environ.* **2017**, *158*, 172–180. [CrossRef]
52. Giraldo, R.; Delicado, P.; Mateu, J. Ordinary kriging for function-valued spatial data. *Environ. Ecol. Stat.* **2011**, *18*, 411–426. [CrossRef]
53. Zhang, L.W.; Zhu, P.; Liew, K.M. Thermal buckling of functionally graded plates using a local Kriging meshless method. *Compos. Struct.* **2014**, *108*, 472–492. [CrossRef]
54. Addesso, P.; Longo, M.; Montone, R.; Restaino, R.; Vivone, G. Interpolation and combination rules for the temporal and spatial enhancement of SEVIRI and MODIS thermal image sequences. *Int. J. Remote. Sens.* **2017**, *38*, 1889–1911. [CrossRef]
55. Fu, D.; Xia, X.; Wang, J.; Zhang, X.; Li, X.; Liu, J. Synergy of AERONET and MODIS AOD products in the estimation of PM2.5 concentrations in Beijing. *Sci. Rep. UK* **2018**, *8*, 10174. [CrossRef] [PubMed]
56. Song, Z.; Fu, D.; Zhang, X.; Wu, Y.; Xia, X.; He, J.; Han, X.; Zhang, R.; Che, H. Diurnal and seasonal variability of PM2.5 and AOD in North China plain: Comparison of MERRA-2 products and ground measurements. *Atmos. Environ.* **2018**, *191*, 70–78. [CrossRef]
57. Trinh, T.T.; Trinh, T.T.; Le, T.T.; Nguyen, T.D.H.; Tu, B.M. Temperature inversion and air pollution relationship, and its effects on human health in Hanoi City, Vietnam. *Environ. Geochem. Health* **2019**, *41*, 929–937. [CrossRef]

58. Kim, S.-W.; Yoon, S.-C.; Kim, J.; Kim, S.-Y. Seasonal and monthly variations of columnar aerosol optical properties over east Asia determined from multi-year MODIS, LIDAR, and AERONET Sun/sky radiometer measurements. *Atmos. Environ.* **2007**, *41*, 1634–1651. [CrossRef]
59. Ju, Y. Tracking the PM2.5 inventories embodied in the trade among China, Japan and Korea. *J. Econ. Issues* **2017**, *6*, 27. [CrossRef]

Publisher's Note: MDPI stays neutral with regard to jurisdictional claims in published maps and institutional affiliations.

Article

Remote Sensing Image Scene Classification with Noisy Label Distillation

Rui Zhang [1,2], Zhenghao Chen [1], Sanxing Zhang [1,2], Fei Song [1,3], Gang Zhang [1], Quancheng Zhou [1,2] and Tao Lei [1,*]

1 Institute of Optics and Electronics, Chinese Academy of Sciences, Chengdu 610209, China; zhangrui182@mails.ucas.ac.cn (R.Z.); plusczh@gmail.com (Z.C.); zhangsanxing18@mails.ucas.ac.cn (S.Z.); sfei_work@ynnu.edu.cn (F.S.); zhanggang@ioe.ac.cn (G.Z.); zhouquancheng18@mails.ucas.ac.cn (Q.Z.)

2 University of Chinese Academy of Sciences, Beijing 100049, China

3 School of Information and Communication Engineering, University of Electronic Science and Technology of China, Chengdu 611731, China

* Correspondence: taoleiyan@ioe.ac.cn

Received: 12 June 2020; Accepted: 19 July 2020; Published: 24 July 2020

Abstract: The widespread applications of remote sensing image scene classification-based Convolutional Neural Networks (CNNs) are severely affected by the lack of large-scale datasets with clean annotations. Data crawled from the Internet or other sources allows for the most rapid expansion of existing datasets at a low-cost. However, directly training on such an expanded dataset can lead to network overfitting to noisy labels. Traditional methods typically divide this noisy dataset into multiple parts. Each part fine-tunes the network separately to improve performance further. These approaches are inefficient and sometimes even hurt performance. To address these problems, this study proposes a novel noisy label distillation method (NLD) based on the end-to-end teacher-student framework. First, unlike general knowledge distillation methods, NLD does not require pre-training on clean or noisy data. Second, NLD effectively distills knowledge from labels across a full range of noise levels for better performance. In addition, NLD can benefit from a fully clean dataset as a model distillation method to improve the student classifier's performance. NLD is evaluated on three remote sensing image datasets, including UC Merced Land-use, NWPU-RESISC45, AID, in which a variety of noise patterns and noise amounts are injected. Experimental results show that NLD outperforms widely used directly fine-tuning methods and remote sensing pseudo-labeling methods.

Keywords: scene classification; teacher-student; noisy labels; knowledge distillation; remote sensing images

1. Introduction

The optical remote sensing image is a powerful source of geographical information since it contains complex geometrical structures and spatial patterns. In recent decades, the remote sensing community has tried to establish an accurate remote sensing image scene classifier. Recent advances in Convolutional Neural Networks (CNNs) make it possible to identify remote sensing scenes with better performance [1,2]. However, many real-world applications for earth observation require large-scale datasets with clean annotations such as ImageNet [3]. It is costly and time-consuming to collect a large-scale remote sensing dataset with high-quality manual annotations. Lack of annotated data has become a bottleneck for the development of deep learning methods in remote sensing and Earth observation. Moreover, the same bottleneck also exists in many other visual tasks.

To tackle the bottleneck, many studies [4] start with leveraging crowd-sourcing platforms, image search engines, or other automatic labeling methods to collect labeled data for natural image scene classification. For example, the Open Images Dataset V4 [5] contains over 30.1 million image-level labels automatically produced by a classifier and a small percentage of labels are verified by crowd-sourcing platforms. These methods significantly reduce the cost of data labeling, which is valuable for applying deep learning in remote sensing image scene classification. The volume of unlabeled images collected by satellites or drones is growing by a few terabytes each day. Low-cost annotations could facilitate the use of abundant image resources. Hence, some methods [6] generate pseudo labels for unlabeled remote sensing images through semi-supervised learning. However, these labels struggle to provide the same asymptotic properties as supervised learning does in high-data regimes. The labels produced by these approaches contain varying degrees of error, i.e., noise, and the performance of classifiers is highly sensitive to massive label noise. Since most of the automatically generated labels are mismatched, it is challenging for traditional learning methods to work on such datasets.

Training on noisy labeled datasets become essential and has attracted much attention in recent years [7–9]. Furthermore, several approaches learning with noisy labels [10–12] have been explored for remote sensing image analysis tasks. Existing methods based on RGB images with noisy labels usually make a strong assumption that all labels are noisy. These studies mostly work on robust algorithms against noisy labels [13], label cleansing methods finding label errors [14] , or combining them together [15]. It was proven that these classifiers have achieved good accuracy on noisy CIFAR10/100 datasets. However, it is difficult and impractical to apply these complex methods to other areas. For remote sensing image scene classification, some of these methods sometimes do not perform as well as direct training. In real-world applications, datasets usually contain a small fraction of images with clean annotations and large amounts of images with noisy or missing labels. In this case, some approaches [16–18] have produced better performance and practicality on large-scale real-world noisy datasets, such as Clothing1M dataset [8] and Open Images V4 dataset [5]. To the best of our knowledge, there is no existing work for remote sensing image scene classification with minimal extra-human supervision.

This work focuses on augmenting existing human-verified labeled datasets with additional noisy labeled data to improve the performance of remote sensing scene classifiers. A more efficient way is explored to learn knowledge from massive noise, instead of simply mix all data or fine-tuning with labeled images. Inspired by Deep Mutual Learning (DML) [19], this paper proposes a novel noisy label distillation framework called NLD based on teacher-student methodology with a decision network, as given in Figure 1. First, the student and teacher jointly learn from each other. Pre-training is no longer a required process. Second, the teacher distills the knowledge learned from noisy data to facilitate the student to learn from full dataset. NLD can even be applied to completely noise-free datasets. This means that our method can be used in a wide range of remote sensing applications. Third, a decision network derived from [20] is introduced, which is easier to optimize in practice and replace the calculation of the mimicry loss. Considering the lack of public datasets with noisy annotations for remote sensing image scene classification, experiments are conducted to evaluate NLD by injecting a series of noises into well-annotated datasets(e.g., UC Merced Land-use [21], NWPU-RESISC45 [22] and AID [23]).

Our contributions are as follows:

- Noise label is introduced for remote sensing image scene classification with minimal extra-human supervision. In practical applications, it is possible to label millions of images with noisy labels at a low-cost.
- A novel and effective end-to-end framework based on teacher-student model namely NLD is proposed for noisy labels distillation. NLD can effectively boost the performance of remote sensing scene classifiers with massive noisy annotations.
- NLD is effective on completely clean datasets. Thus, NLD can be further extended to model distillation for network compression.

- Pseudo-labeling methods can automatically generate nearly infinite noisy annotated images at no additional cost. The network trained by NLD achieves a better performance than other pseudo-labeling methods.
- Several new practical benchmarks are proposed for remote sensing image scene classification with different types of noisy labels.

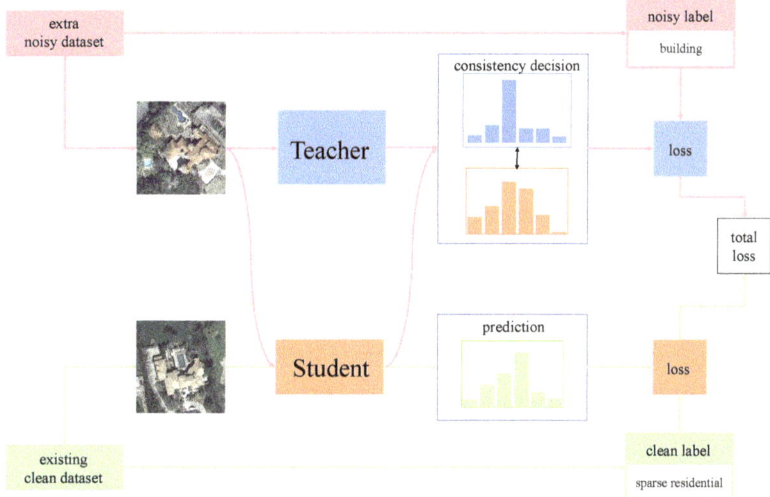

Figure 1. A high-level illustration of NLD. The student and teacher mutually learn knowledge of clean and noisy labels.

This paper is organized as follows: Section 2 introduces the research status and the challenges. Section 3 describes the overall framework of NLD. Section 4 presents the implementation details of experiments and analyzes the result. Finally, Section 5 concludes our paper and gives an outlook.

2. Related Works

In this section, we will briefly review existing related works on remote sensing image scene classification and learning from noisy labels.

2.1. Remote Sensing Image Scene Classification

Remote sensing image scene classification aims to distinguish the semantic category of an image, which is a fundamental problem for understanding high-level geospatial information. With the development of deep learning methods, many CNN architectures (e.g., ResNet [24], VGG [25]) have achieved remarkable performance on many remote sensing public datasets. However, there are large intra-class variations and small inter-class dissimilarities between different remote sensing scenes. These problems will decrease the recognition abilities of models for some categories. To address these challenges, many studies focus on how to learn discriminative feature representations. Nogueira et al. [2] analyzed the use of different networks in the field of remote sensing. Chaib et al. [1] proposed an adequate method for feature fusion and introduced discriminant correlation analysis to represent the fused features. Zhang et al. [26] proposed a newly designed CapsNet to deal with the remote sensing image scene classification problem. Li et al. [27] proposed a unified feature fusion framework based on attention mechanism to improve the classification performance.

These algorithms are all data-driven algorithms, which means large-scale datasets are required in practice. To facilitate the application of these methods to more fields that have little data with clean annotations, NLD can be widely used with various models including the above research.

2.2. Learning from Noisy Labels

Most of methods learning from noisy datasets aim to directly learn without clean labels available. These approaches usually focus on noise-robust algorithms and label cleansing methods. Wang et al. [13] proposed symmetric cross entropy (SCE) loss to boost cross-entropy (CE) symmetrically with a noise-robust counterpart reverse CE. Northcutt et al. [14] proposed confident learning for characterizing, identifying, and learning with noisy labels. Kim et al. [15] proposed Selective Negative Learning and Positive Learning (SelNLPL) to filter and learn with noisy data. These methods face the problem of discriminating difficulty from mismatched labels.

Our approach belongs to a practical stream, assuming that both clean and noisy labels of the dataset are known [8,28]. This is a more practical scenario, allowing researchers to focus on leveraging noisy labeled data to enhance existing fully supervised algorithms. Veit et al. [16] proposed a learning approach for multi-label image classification using clean labeling combined with massive noise labeling. Hu et al. [18] proposed a method to automatically identify credible annotations in the massive noisy labels under weakly supervised learning. Many semi-supervised learning algorithms, especially pseudo-labeling algorithms, can also be categorized into such scenarios [29]. Han et al. [6] proposed a framework based on deep learning features, self-labeling techniques and decision evaluation methods under semi-supervision for remote sensing image scene classification and annotating datasets. The works closer to ours comes from Li et al. [17] and Li et al. [30].To achieve noisy label learning, they proposed a teacher-student framework, which comes from knowledge distillation [31]. To take full use of the whole data space, traditional knowledge distillation and many other similar noise-robust methods use the student model to mimic the large pre-trained teacher model by providing training experiences. These experiences are called "dark knowledge".

In practice, a smaller network with the same precision is needed because of the cost, i.e., a student network. However, due to the existence of noisy labels, even under the guidance or regularization of a powerful network pre-trained with clean data, small networks are still prone to overfit to noisy labels. This may even lose the knowledge of the original clean data.

3. Method

3.1. Problem Formulation

Our goal is to train a remote sensing scenes classifier using a dataset with automatically collected noise labels and a part of human-verified clean labels available. The source of noisy labels may come from collects from the web or predictions from models trained on clean data or other ways. Furthermore, the framework can be used for large-scale datasets with fully clean annotations to improve the performance of networks under traditional supervised learning.

Formally, we define the notations for our study. Let $\mathcal{D} = \mathcal{D}_c \bigcup \mathcal{D}_n$ donates the entire large training dataset, where $\mathcal{D}_c$ is the clean subset and $\mathcal{D}_n$ is the remaining noisy subset. In a single label classification problem, $\mathcal{D}_c = \{(\vec{x}_i, y_i) | \, i = 1, 2, \cdots, N_c\}$ and $\mathcal{D}_n = \{(\vec{x}_j, y_j) | \, j = 1, 2, \cdots, N_n\}$, which contains N_c and N_n samples from M classes, respectively; $y_i \in \{1, 2, \ldots, M\}$ and $y_j \in \{1, 2, \ldots, M\}$ donate the label corresponding to image $\vec{x}_i$ and $\vec{x}_j$. In this work, the ratio of $\mathcal{D}_n$ to $\mathcal{D}_c$ is not limited, because NLD can improve the performance of classifiers in different practical applications.

As shown in Figure 2, NLD is formulated with a cohort of two classifiers g and h. The classifier g is the large teacher model that is used to distill and transfer the knowledge of noise. In addition, its backbone is a powerful network such as a ResNet-50 [24]. The student model h is designed to learn from the clean labels and guided the learning process by the knowledge of noise which is distilled from the teacher network T. The network S is a network that is same as or shallower than network

T (e.g., ResNet-34 [24] and VGG-16 [25]). The logits $\vec{r_1}$ for $\vec{x_j}$ given by the teacher network T can be represented as

$$\vec{r_1} = \mathcal{F}_n\left(\vec{x_j}\right), \tag{1}$$

where the $\mathcal{F}_n$ is a nonlinear transformation in teacher network T. Similarly, the logits $\vec{r_2}$ and $\vec{c_1}$ can be represented as

$$\vec{r_2} = \mathcal{F}_c\left(\vec{x_j}\right), \tag{2}$$

$$\vec{c_1} = \mathcal{F}_c\left(\vec{x_i}\right), \tag{3}$$

where the $\mathcal{F}_c$ is a nonlinear transformation in student network S.

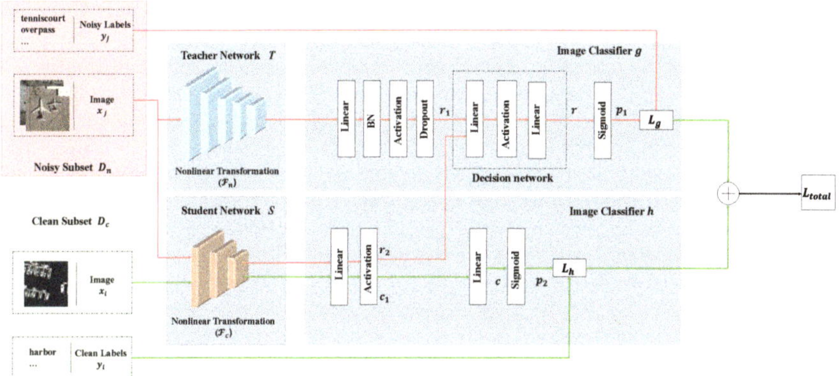

Figure 2. The overview of the proposed framework to train a remote sensing scenes classifier from a large dataset $\mathcal{D}_n$ with noisy labels and a small dataset $\mathcal{D}_c$ with manually verified labels. The framework consists of teacher network T, student network S, decision network, fully connected layers, and predictor of softmax. In the training phase, two loss terms L_g and L_h (a CE loss with noisy labels and a CE loss with clean labels) are minimized jointly. The teacher model T transfers the "dark knowledge" distilled from noisy subset to the student model S through the decision network. In the inference phase, a classifier containing the student network S, fully connected layers and softmax can give the correct predictions.

For classifier g and h, the supervision depends on the source of the training sample. For image $\vec{x_j}$ from the noisy dataset $\mathcal{D}_n$, the classifier g is supervised by the noisy label y_j. For sample $\vec{x_i}$ from the clean dataset $\mathcal{D}_c$, supervision comes directly from the verified label y_i.

3.2. Noisy Distillation

In contrast to the previous work on teacher-student models including [17,30], we need to pre-train a teacher model with a small part of or the entire dataset: the teacher model and student model are trained together to learn latent noisy label distributions to improve the performance of student network supervised with the clean subset. NLD is motivated by DML which leverages a teacher-student framework to improve the representation of the network. The details will be analyzed in the later part of this section.

The student network learns the knowledge of clean data and acquires the distilled knowledge of the noisy dataset. The teacher takes advantage of powerful deep network architectures to learn features of noisy labels at various levels of abstraction rather than simply memorizing these. Besides, noise knowledge is distilled by comparing the outputs of the student and teacher simultaneously. To that end, the student and teacher model are trained by a mutual learning approach which originates from knowledge distillation. Noted that NLD is different from DML and other similar approaches. To match noisy label distributions, a metric between two branch's representation vectors $\vec{r_1}$ and $\vec{r_2}$ needs to be defined. As a loss function, Kullback Leibler (KL) Divergence is the most widely used. The KL distance from $\vec{r_1}$ and $\vec{r_2}$ is computed as

$$D_{KL}\left(\vec{r_2} \parallel \vec{r_1}\right) = \sum_{i=1}^{N_n} \sum_{m}^{M} r_2^m\left(\vec{x_j}\right) \log \frac{r_2^m\left(\vec{x_j}\right)}{r_1^m\left(\vec{x_j}\right)}, \tag{4}$$

where the r_1^m is the score of class m in logits $\vec{r_1}$ and the r_2^m is the score of class m in logits $\vec{r_2}$.

According to the formula, KL divergence is asymmetric. Hence, the KL distance between the two networks is different. One can instead use a symmetric KL-divergence such as

$$D_{SKL} = D_{KL}\left(\vec{r_2} \parallel \vec{r_1}\right) + D_{KL}\left(\vec{r_1} \parallel \vec{r_2}\right). \tag{5}$$

Compared to teacher network T, student network S has similar representation capacities, but it is harder to learn appropriate parameters. In DML and other similar knowledge distillation algorithms, both teacher network and student network are trained on clean datasets. These studies expect the student network to mimic the classification probabilities and feature representations of the teacher network. The objective functions of the two networks are the same. Therefore, a simple combination of CE loss and KL divergence can facilitate a better student network from the entire clean dataset. However, how to combine and optimize these two different kinds of losses will be a difficult problem in our tasks. Our teacher network T is supervised by noise labels and our student network S is supervised by clean labels. The student network S should not totally mimic outputs of the teacher network T. By imitating and comparing, the purpose is to distill the knowledge from the noisy dataset, which is the intersection of clean student's features and noisy teacher's features. In the meanwhile, as mentioned above, a simple combination of CE loss and KL divergence would work on two networks identical to each other. Although this can be changed by adding some weights before the combination, there are too many options for hyper-parameters.

To address these problems, NLD feeds outputs of the two networks simultaneously into a decision network derived from [20]. The decision network simply consists of fully connected layers with a single output. In [20], this network is used to measure the similarity between two different images with siamese network. As discussed above, NLD has different settings from images similarity measurement methods. Different logits of two same image patches are mapping from different networks. Furthermore, the similarity of two networks is measured through the decision network. In addition, the decision network has learnable parameters. Instead of relying on the combination of different loss functions with hyper-parameters, this can automatically learn weights that fit the noisy label knowledge distillation. Because the original logits are mapping from the same image, the output r of decision network is still the original image feature mapping. The probability of class m for sample $\vec{x_j}$ given by decision network is computed as

$$p_1^m\left(\vec{x_j}\right) = \frac{\exp\left(r^m\right)}{\sum_{m=1}^{M} \exp\left(r^m\right)}. \tag{6}$$

Subsequently, the classifier g is supervised by noisy labels and the classifier h is supervised by clean labels. In this way, the student network can learn clean knowledge and similar knowledge between clean labels and noise labels, i.e., noise distillation. At the same time, NLD does not need a mimicry loss, so training is faster and more flexible than traditional distillation methods. In the meanwhile, the decision network also increases inference time as it requires combinations of two vectors. However, our goal is to train a student network guided by the teacher network. Therefore, only the student network is used for testing, while the decision network is not used.

3.3. Model Training

In original knowledge distillation and DML, the whole objective function consists of a supervised loss (e.g., CE loss) and a mimicry loss(e.g., KL divergence). In contrast, CE loss is used as the supervised loss for classifier g and h, respectively. In addition, they can be rewritten as:

$$L_g = -\sum_{j=1}^{N_n} y_j \log(p_1), \quad (7)$$

$$L_h = -\sum_{i=1}^{N_c} y_i \log(p_2), \quad (8)$$

where L_g and L_h are the losses for the corresponding classifier g and h, respectively. Given the above definitions, the overall loss for the proposed model is constructed by two losses as follows:

$$L_{total} = \alpha L_h + \beta L_g, \quad (9)$$

where α and β denote weight factors that need to be set based on student network, teacher network and noisy dataset.

Training a network with a noisy dataset can lead the network to memorize these noises. To avoid the teacher network overfitting on noisy data, which will deteriorate the performance of noise distillation and may even mislead the student to have exploding gradients, batch normalization(BN) [32] and dropout layer [33] with a constant probability of 0.6 are applied between the teacher network and the decision network.

3.4. Extension to Pseudo-Labeling

Semi-supervised learning requires a small amount of manually labeled clean data, which is consistent with NLD. However, semi-supervised learning datasets usually contain a small amount of labeled data and a large amount of unlabeled data. Because NLD does not use additional mimicry loss, unlabeled data cannot be used directly. Pseudo-labeling belongs to the self-learning scenario which is often used in semi-supervised learning. Under the self-training settings, pseudo-labels are obtained by predicting unlabeled data through the models trained on labeled data. Some of the pseudo-labels will be mislabeled. These data with the pseudo-labels can be treated as a large noisy dataset and NLD can extend to semi-supervised learning.

Following [6], the pseudo-labeling method used is illustrated in Figure 3, which is close to traditional co-training. Denote the labelled and unlabeled subsets as $\mathcal{D}_l$ and $\mathcal{D}_u$, where the entire training dataset is $\mathcal{D}_s = \mathcal{D}_l \bigcup \mathcal{D}_u$. First, there are two different classifiers f_1 and f_2 trained on the small labeled dataset $\mathcal{D}_l$, respectively. Given a batch of unlabeled images $\vec{x}' \in \mathcal{D}_u$, two predictions $\widetilde{y_1}$ and $\widetilde{y_2}$ are provided by the classifiers f_1 and f_2. Then, $\widetilde{y_1}$ and $\widetilde{y_2}$ can be represented as

$$\widetilde{y_1} = f_1\left(\vec{x}'\right), \quad (10)$$

$$\widetilde{y_2} = f_2\left(\vec{x}'\right). \quad (11)$$

Only when $\widetilde{y_1} = \widetilde{y_2}$, the predictions of the classifiers f_1 and f_2 will be regarded as the pseudo-label y' corresponding to $\vec{x}'$, and other different results will be discarded. Apparently, this process will reduce the dataset size from $\mathcal{D}_u$, which typically affects the final performance. In fact, it removes low confidence predictions from pseudo-labels and reduces the noise level of the labels. High-quality pseudo-labels can improve performance and the robustness of the model. Furthermore, it does not need to choose a confidence threshold or manual selection. This is a more efficient pseudo-labeling method.

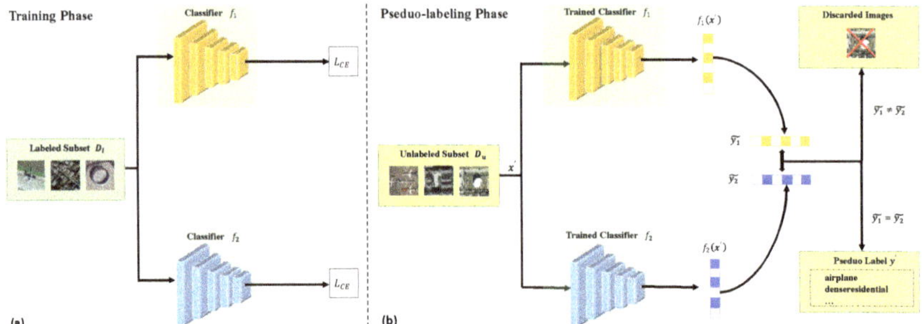

Figure 3. Illustration of the pseudo-labeling method, which includes two phases: training two classifiers and pseudo-labeling. (**a**) Two different classifiers f_1 and f_2 trained on the small manually labeled subset $\mathcal{D}_l$, respectively. They provide two views of the data. (**b**) The trained models can predict labels on a batch of unlabeled data. When the inferences are the same, the predicted labels will remain as pseudo-labels for the corresponding images, and the rest will be discarded. L_{CE} donate CE loss. $\vec{y_1}$ and $\vec{y_2}$ represent the predictions of two classifiers, respectively. $\vec{y}'$ indicates pseudo-labels of the batch of images $\vec{x}'$.

4. Experiments

In this section, we explain how to construct the mimic noisy datasets and describe the experimental details of our comparison with other methods on these datasets and evaluate NLD.

4.1. Datasets and Settings

4.1.1. Datasets

UC Merced Land-use dataset is a classical land-use dataset, which contains 21 different scenes and 2100 images. Each image has 256 × 256 pixels and high-resolution in RGB color space with a spatial resolution of 0.3 m. They were all manually extracted from the USGS National Map Urban Area Imagery Collection.

NWPU-RESISC45 dataset has a total number of 45 scene classes and 700 images with a size of 256 × 256 for each class. Most of the images are middle to high spatial resolution, which varies from 30 m to 0.2 m. They are all cropped from Google Earth. The dataset takes eight popular classes from UC Merced Land-use dataset and some widely used scene categories from other datasets and research.

AID is a large-scale aerial image dataset with 30 aerial scene types. The dataset is composed of 10,000 images which are multi-resolution and multi-source. The size of each image is fixed to be 600 × 600. The number of images in each class is imbalanced. This dataset is challenging because of the large intra-class diversities.

These datasets have many overlapped classes (e.g., sparse residential, medium residential and dense residential) that can easily confuse non-expert. It is particularly challenging for computer vision researchers with little geography knowledge to label such a dataset manually. As for crowd-sourcing or automatic labeling, it will be more prone to make errors. Actually, based on the existing public datasets, when we need to use them in real-world applications, additional data will be used. Only experts can avoid label noise, which is expensive.

Experiments are conducted on these three datasets. In addition, as shown in Table 1, each dataset is randomly split into 60% training subset, 20% validation subset and 20% test subset. Because the existing datasets lack noisy labels, simulated approaches are taken to evaluate NLD. Three different types of noise are injected into the split training set of all the three datasets separately.

Table 1. Sample sizes for different datasets.

Datasets	Entire Dataset	Training Subset	Validation Subset	Test Subset
UC Merced Land-use	2100	1260	420	420
NWPU-RESISC45	31,500	18,900	6300	6300
AID	10,000	6000	2000	2000

Symmetric noise: The symmetric noise is a type of uniform noise, which is generated by a random label among the classes to replace the ground-truth label with equal probabilities. This type of noisy subset represents an almost zero-cost annotation method, which means there are many unlabeled images, and labels are labeled in a completely random way. Experiments on this noise can prove that, through NLD, this labeling method is also feasible in some extremely low-cost scenarios.

Asymmetric noise: This type of noise is class dependent noise and it mimics some of the real-world noise for visually similar and semantically similar categories.

For UC Merced Land-use, to the best of our knowledge, there is no related noise label mapping method before. After observing the features of images and division of scene classes, asymmetric noise was generated by mapping *chaparral* $\rightarrow$ *agricultural*, *runway* $\leftrightarrow$ *airplane*, *tennis court* $\rightarrow$ *baseball diamond*, *river* $\rightarrow$ *beach*, *mobile home park* $\rightarrow$ *parking lot*, *freeway* $\leftrightarrow$ *overpass*, *sparse residential* $\rightarrow$ *buildings*, *harbor* $\rightarrow$ *mobile home park*, *medium residential* $\leftrightarrow$ *dense residential* as shown in Figure 4.

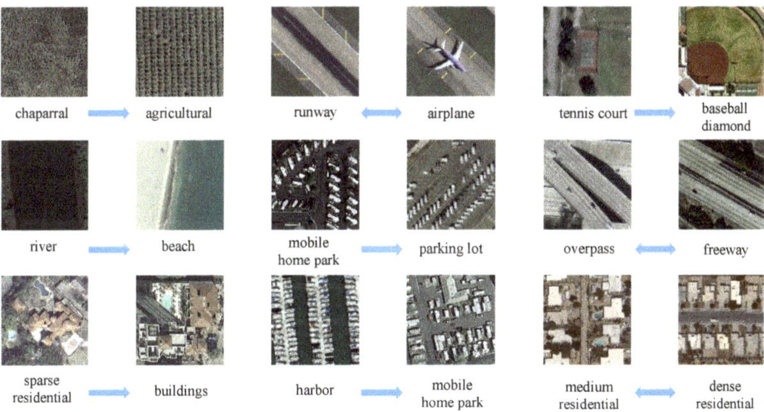

Figure 4. Examples of asymmetric noise mapping scenes in the UC Merced Land-use dataset.

For NWPU-RESISC45, *baseball diamond* $\rightarrow$ *medium residential*, *beach* $\rightarrow$ *river*, *dense residential* $\leftrightarrow$ *medium residential*, *intersection* $\rightarrow$ *freeway*, *mobile home park* $\leftrightarrow$ *dense residential*, *overpass* $\leftrightarrow$ *intersection*, *tennis court* $\rightarrow$ *medium residential*, *runway* $\rightarrow$ *freeway*, *thermal power station* $\rightarrow$ *cloud*, *wetland* $\rightarrow$ *lake*, *rectangular farm land* $\rightarrow$ *meadow*, *church* $\rightarrow$ *palace*, *commercial area* $\rightarrow$ *dense residential* are mapped, following [12]. Figure 5 shows representative images in this dataset.

For AID, the classes are flipped by mapping *bareland* $\leftrightarrow$ *desert*; *center* $\rightarrow$ *storage tank*; *church* $\rightarrow$ *center*, *storage tank*; *dense residential* $\leftrightarrow$ *medium residential*; *industrial* $\rightarrow$ *medium residential*; *meadow* $\rightarrow$ *farm land*; *play ground* $\rightarrow$ *meadow*, *school*; *resort* $\rightarrow$ *medium residential*; *school* $\rightarrow$ *medium residential*, *play ground*; *stadium* $\rightarrow$ *play ground*, following [12]. Figure 6 shows examples from this dataset.

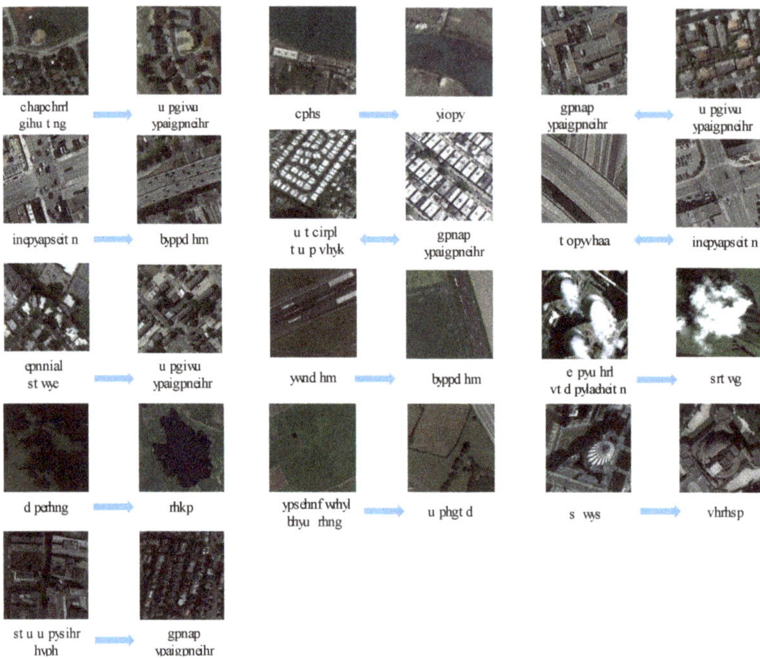

Figure 5. Examples of asymmetric noise mapping scenes in the NWPU-RESISC45 dataset.

Figure 6. Examples of asymmetric noise mapping scenes in the AID dataset.

Pseudo-Labeling noise: Pseudo-labeling methods can assign labels to unlabeled images automatically, which can reduce costs. However, there are not completely correct pseudo-labels.

To ensure a fair comparison, following the idea of SSGA-E [6], the full training set is randomly divided into six parts and randomly select one of them as a small clean subset. Then, two different classifiers are trained on the small clean subset and make pseudo labels for the rest of the train set. In SSGA-E [6], two networks are ResNet-50 and VGG-S [34], respectively. However, VGG-S is rarely used in practice, which can cause many problems in deployment. As a result, VGG-S is replaced with the VGG-19 [25], which has lower accuracy but is more widely used. These unlabeled subsets with automatically generated labels can be viewed as the noisy subset. In addition, since this method does not label all images, some of the uncertain images are removed from the subset and the noise subset will be smaller than the original subset. The number of annotations obtained for unlabeled images of different datasets is listed in Table 2.

Table 2. Number of samples contained in different subsets. The unlabeled subset is $\frac{5}{6}$ of the entire training set. Pseudo-labeled subset is generated in unlabeled subset by the automatic labeling method trained with the clean labeled subset (i.e., $\frac{1}{6}$ of the entire training set) as a clean subset.

Datasets	Entire Training Subset	Clean Labeled Subset	Unlabeled Subset	Pseudo-Labeled Subset
UC Merced Land-use	1260	210	1050	859
NWPU-RESISC45	18,900	3150	15,750	13,625
AID	6000	1000	5000	4535

4.1.2. Baselines and Model Variants

To evaluate the performance improvement of NLD, our approach is compared with some pseudo-labeling methods [6]. Several related baselines are also provided for symmetric noise, asymmetric noise. In addition, NLD is used as the base model for some other variants to verify the effectiveness of NLD. The details of the baselines and variants are as follows.

Baseline-Clean: A backbone network of the student model is trained for remote sensing scenes classification using the clean subset. This can be regarded as the lower bound of NLD. Our method uses the noisy subset to improve performance on this baseline.

Baseline-Noise: A backbone network of the student model is trained solely on noisy labels from the training set. This baseline can be viewed as a measurement of the quality of noisy labels.

Baseline-Mix: A backbone network of the student model is trained using a mix of clean and noisy labels with standard CE loss. This baseline shows the damage caused by noisy subsets.

SCE Loss: Under the supervision of SCE loss, a model is trained on the entire dataset with both clean and noisy labels. Parameters for SCE are configured as $\alpha = 0.1$ and $\beta = 1.0$.This is a baseline for a noise-robust method.

Noise model fine-tune with clean labels (Clean-FT): It is a common approach, which uses the clean subset directly to fine-tune the whole network of Baseline-Noise. This method is prone to overfit if there are few clean samples.

Noise model fine-tune with mix of clean and noisy labels (Mix-FT): To address the problem caused by limited clean labels, fine-tuning the Baseline-Noise with mixed data is also a common approach.

NLD with CE loss (NLD): NLD is trained on both the original clean datasets and different noisy ratios of datasets. For a completely clean dataset, one image is used as input simultaneously for the teacher and student, which is close to DML.

4.1.3. Experimental Settings

All experiments are implemented with PyTorch framework [35] and conducted on an NVIDIA GeForce Titan X GPU. The networks used in our experiments are shown in Table 3. These networks are all pre-trained on ImageNet. Although VGG architecture has a larger number of parameters and needs more floating point operations(FLOPs), ResNet architecture has stronger feature representation capabilities-based residual modules. Therefore, teacher networks in all experiments are ResNet architecture. For UC Merced Land-use dataset, it is worth mentioning that SSGA-E [6] uses VGG-S and VGG-16, but after our experiments, the network with VGG architectures will be over-fitting because the size of this dataset is small. So the actual network used is modified VGG architectures with BN to learn this dataset. As a preprocessing step, random flip, random gaussian blur and resize images to 224×224 are used. For optimization, we use Adam with weight decay of 10^{-2}, batch size of 32 and initial learning rate of 10^{-4}. The leaning rate will decrease according to the exponential decay with the multiplicative factor of 0.98 in each epoch. All networks mentioned in Section 4.1.2 are trained for 200 epochs. Besides, for NLD, a batch of images is half clean and half noise. In general, the weight factors are set to $\alpha = 10$ and $\beta = 2$. For additional experiments, experiments are conducted with more different factors, losses and networks, which will be detailed in Section 4.6

Table 3. Comparison of various network architecture.

Network Type	Million Parameters	GFLOPs
ResNet-34	21.819	3.679
ResNet-50	25.578	4.136
VGG-16	138.379	15.608
VGG-16 with BN	138.387	15.662
VGG-19	143.688	19.771
VGG-19 with BN	143.699	19.830

4.2. Results on UC Merced Land-Use

The results on the original UC Merced Land-use without any label noise and the UC Merced Land-use with symmetric label noise are reported in Table 4. Two confusion matrices for noise-free UC Merced Land-use are shown in Figures 7 and 8, respectively. It is noticeable that the student network (ResNet-34) can significantly benefit for NLD when learning from the original noise-free dataset. Therefore, NLD can also be regarded as a model distillation-like process, without additional data and pre-trained models. For symmetric noise, this type of noise label is completely random and there is little correct information for NLD distilling the knowledge in the noisy subset. Our method can still make better performance and robustness of the student network in most cases. As for $D_c : D_n = 8 : 2$ and $D_c : D_n = 2 : 8$ cases, it revealed that when the clean subset D_c or noisy subset D_n is too small (e.g., 252 samples), clean labels or randomly generated labels are too weak to bootstrap the performance. Instead of improving performance, other common approaches even hurt the performance. When the label quality of the noise subset is extremely low, a lot of error guidance will be provided. Specifically, different fine-tuning methods require a pre-trained model of the noise subset, which may get worse initialization values than the ImageNet [3] pre-trained model. If the two subsets are mixed, the noise labels will become adversarial examples, which confuse the network. SCE or other noise-robust methods can alleviate this problem, but the performance is still far from the method with a small number of clean labels available.

Table 5 shows the results for asymmetric label noise. This noise is closer to the real scene, similar to crowd-sourcing labeling or crawling data from Internet. According to the results of Baseline-Noise, such labels can provide a more valuable pre-trained model than labels with symmetric noise. Clean-FT and Mix-FT provide clear improvements compared to Baseline-Clean and Baseline-Mix, respectively. However, for mix-based methods, during training, the learning process of the model on the clean subset will be continuously misguided by the noise labels. As the noise ratio increases and clean ratio decreases, less clean data is difficult to fight against more noisy data, the performance of Mix-FT and SCE Loss is severely impaired. For NLD, the framework can maintain a better performance with fewer clean labels and more noisy labels. When $D_c : D_n$ goes from $2 : 8$ to $8 : 2$, the performance of the model will only decrease by 1.62%. It is particularly noteworthy that when $D_c : D_n = 2 : 8$, NLD can exceed 6.67% of the Baseline-Clean.

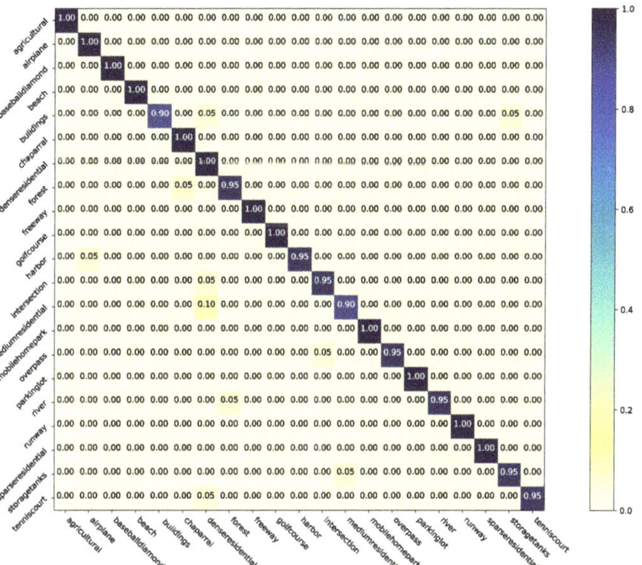

Figure 7. The confusion matrix of Baseline-Clean with full UC Merced Land-use dataset.

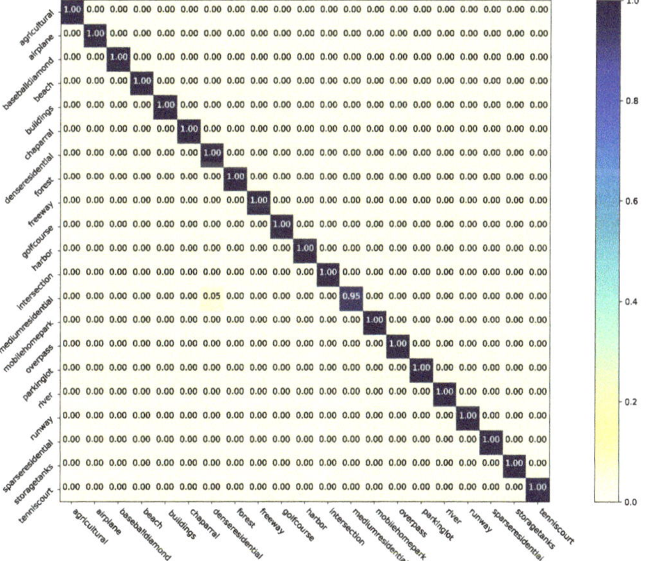

Figure 8. The confusion matrix of NLD with full UC Merced Land-use dataset.

Table 4. Classification accuracy (%) on the UC Merced Land-use test set for different methods trained with the original noise-free dataset and symmetric label noise. We report the mean and standard error across 5 runs.

Methods	Network Types	None	Symm			
		$D_c : D_n$	$D_c : D_n$			
		10 : 0	8 : 2	6 : 4	4 : 6	2 : 8
Baseline-Clean	ResNet-34	98.66 ± 0.84	98.48 ± 0.75	96.52 ± 1.27	94.86 ± 0.89	89.14 ± 1.01
Baseline-Noise	ResNet-34	-	4.86 ± 1.65	6.05 ± 1.86	5.14 ± 0.76	6.14 ± 1.23
Baseline-Mix	ResNet-34	-	91.98 ± 1.55	83.66 ± 1.96	69.67 ± 2.30	43.10 ± 2.40
SCE Loss	ResNet-34	-	91.09 ± 0.69	81.43 ± 1.37	70.86 ± 3.81	44.67 ± 2.91
Mix-FT	ResNet-34	-	91.95 ± 1.11	76.81 ± 2.33	56.19 ± 1.93	29.90 ± 1.88
Clean-FT	ResNet-34	-	98.38 ± 0.38	97.29 ± 0.47	94.14 ± 0.75	87.10 ± 1.15
NLD	ResNet-50+ResNet-34	**99.08 ± 0.40**	**98.86 ± 0.28**	**97.43 ± 0.63**	**95.86 ± 0.29**	**89.28 ± 0.42**

Table 5. Classification accuracy (%) on the UC Merced Land-use test set for different methods trained with asymmetric label noise. We report the mean and standard error across 5 runs.

Methods	Network Types	Asym			
		$\mathcal{D}_c : \mathcal{D}_n$			
		8 : 2	6 : 4	4 : 6	2 : 8
Baseline-Clean	ResNet-34	98.14 ± 0.65	97.09 ± 0.63	94.62 ± 1.57	90.71 ± 1.23
Baseline-Noise	ResNet-34	42.67 ± 0.41	43.53 ± 0.82	43.95 ± 0.32	43.23 ± 0.58
Baseline-Mix	ResNet-34	90.76 ± 0.84	78.95 ± 2.10	65.57 ± 1.74	54.29 ± 1.49
SCE Loss	ResNet-34	90.67 ± 0.73	81.48 ± 2.53	66.76 ± 2.91	54.08 ± 0.50
Mix-FT	ResNet-34	89.96 ± 1.25	79.86 ± 2.62	67.14 ± 1.20	54.95 ± 1.82
Clean-FT	ResNet-34	98.33 ± 0.54	96.62 ± 1.05	95.57 ± 0.89	92.67 ± 1.53
NLD	ResNet-50+ResNet-34	**99.00 ± 0.18**	**97.95 ± 0.44**	**97.57 ± 0.49**	**97.38 ± 0.56**

4.3. Results on NWPU-RESISC45

In this experiment, NLD is tested on NWPU-RESISC45 with different noisy types. Table 6 summarizes the classification accuracy (%) of ResNet-34 trained with/without NLD. According to Baseline-Noise, asymmetric noise can provide more correct information due to the larger scale of NWPU-RESISC45 than UC Merced Land-use. Thus, Clean-FT can benefit from asymmetric noisy labels. However, the performance of other methods is still compromised by the noise. On the contrary, NLD has strong robustness and can benefit from different ratios and types of noisy labels. As for the test set accuracy, NLD has clearly improved the baseline and direct fine-tuning. Figures 9 and 10 show the confusion matrices of NLD and Baseline-Clean on original NWPU-RESISC45. It can be observed that NLD improves the performance of the student network as a method of model distillation.

Table 6. Classification accuracy (%) on the NWPU-RESISC45 test set for different methods.

Methods	Network Types	None	Symm		Asym	
		$D_c : D_n$	$D_c : D_n$		$D_c : D_n$	
		10 : 0	6 : 4	4 : 6	4 : 6	2 : 8
Baseline-Clean	ResNet-34	94.95	91.89	90.14	90.32	90.05
Baseline-Noise	ResNet-34	-	3.59	3.49	65.13	65.08
Mix-FT	ResNet-34	-	35.03	21.27	75.97	68.83
Clean-FT	ResNet-34	-	87.43	84.46	91.95	91.29
Ours	ResNet-50+ResNet-34	**95.86**	**93.79**	92.81	**95.76**	**94.59**

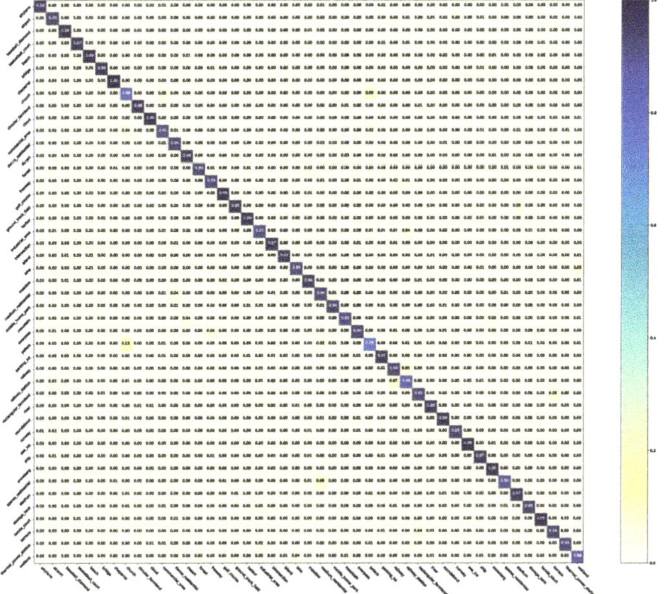

Figure 9. The confusion matrix of Baseline-Clean for the NWPU-RESISC45.

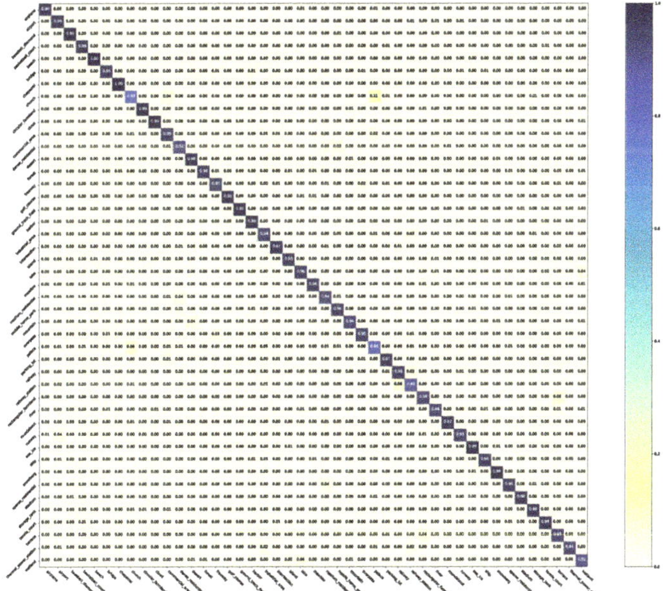

Figure 10. The confusion matrix of NLD for the NWPU-RESISC45.

4.4. Results on AID

Next, the performance of NLD is evaluated on the AID dataset. Table 7 shows the results. As the classes of AID are imbalanced, it is more challenge using noise labels. It can be observed that all methods are significantly affected by symmetric noise, especially when the noise rate increases. In contrast, asymmetric noise can change the imbalance of the data distribution. As a result, NLD can benefit from asymmetric noisy labels and improve performance. The gap between NLD and Clean-Baseline became especially apparent when the noise rate increased to larger values. Our method can be applied to scenarios with more noisy labels. For example, when the asymmetric noise rate is 2 : 8, NLD obtains 2.3% higher accuracy than Baseline-Clean and 3.35% higher than Clean-FT. The confusion matrices for the AID dataset with asymmetric noise of $D_c : D_n = 2 : 8$ are shown in Figures 11 and 12. The results of NLD are significantly better than Baseline-Clean.

Table 7. Classification accuracy (%) on the AID test set for different methods.

Methods	Network Types	None	Symm		Asym	
		$D_c : D_n$	$D_c : D_n$		$D_c : D_n$	
		10 : 0	**6 : 4**	**4 : 6**	**4 : 6**	**2 : 8**
Baseline-Clean	ResNet-34	96.30	95.70	**94.95**	95.10	92.95
Baseline-Noise	ResNet-34	-	6.85	4.7	59.62	59.57
Mix-FT	ResNet-34	-	20.99	11.49	77.71	68.77
Clean-FT	ResNet-34	-	83.96	12.39	94.15	91.90
NLD	ResNet-50+ResNet-34	**96.35**	**95.70**	93.60	**95.90**	**95.25**

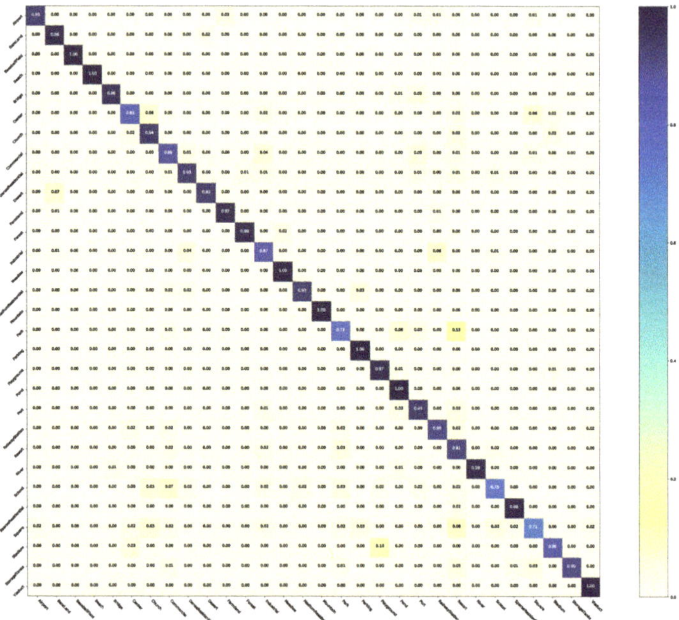

Figure 11. The confusion matrix of Baseline-Clean for the AID dataset with asymmetric noise of $D_c : D_n = 2 : 8$.

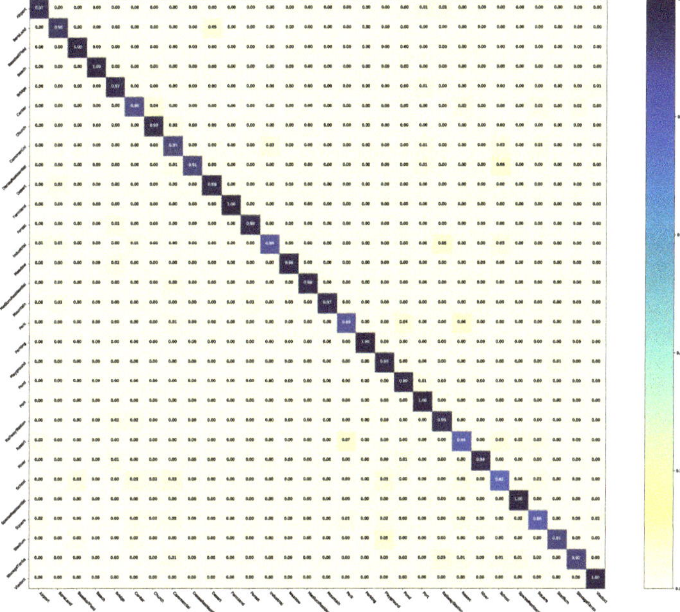

Figure 12. The confusion matrix of NLD for the AID dataset with asymmetric noise of $D_c : D_n = 2 : 8$.

4.5. Comparison with Pseudo-Labeling

We explore pseudo-labeling in UC Merced Land-use, NWPU-RESISC45 and AID. For all datasets, one-sixth of the training images per class is randomly selected as labeled, and the rest of images is treated as unlabeled. Experiments are compared with three pseudo-labeling strategies: (1) traditional self-training with single network; (2) traditional co-training with two networks respectively; (3) SSGA-E [6] with three networks.

Tables 8 and 9 shows the result from Han et al. [6], supplemented with our results. NLD achieves the best overall accuracy in all cases. For the UC Merced Land-use, Resnet34 is more effective as a student network when there is less unlabeled data. When leveraging entire unlabeled subset, VGG-16 shows better performance as a student network. With a larger scale of labeled data (e.g., NWPU-RESISC45), the improvement of our framework is higher. This confirms that NLD benefits pseudo-labeling scenarios.

Table 8. The effect of the unlabeled sample ratio on accuracy for the UC Merced Land-use test set reported by Han et al. [6], supplemented with our results.

Methods	Network Types	Unlabeled Samples				
		210	420	630	840	1050
Self-training [6]	VGG-S			-		86.14 ± 1.87
	ResNet50					91.57 ± 2.00
Co-training [6]	ResNet50&&VGG-S	89.75 ± 1.27	91.62 ± 0.93	92.58 ± 0.78	93.42 ± 1.32	93.75 ± 1.42
SSGA-E [6]	ResNet50&&VGG-S+VGG16	91.42 ± 0.95	92.68 ± 0.87	93.56 ± 1.42	94.21 ± 1.18	94.52 ± 1.38
NLD	ResNet50&&VGG-19+ResNet50+VGG16	91.48 ± 0.80	92.10 ± 0.52	92.67 ± 0.74	93.00 ± 0.82	**95.15 ± 0.85**
NLD	ResNet50&&VGG-19+ResNet50+ResNet34	**93.43 ± 0.55**	**94.19 ± 0.71**	**94.81 ± 0.46**	**94.52 ± 0.87**	93.86 ± 0.99

Table 9. Comparison with results on the NWPU-RESISC45 and AID test set reported by Han et al. [6].

Methods	Network Types	Dataset	
		NWPU-RESISC45	**AID**
Self-training [6]	VGG-S	81.46	86.02
	ResNet-50	85.82	89.38
Co-training [6]	ResNet-50&&VGG-S	87.25	90.87
SSGA-E [6]	ResNet-50&&VGG-S+VGG-16	88.60	91.35
NLD	ResNet-50&&VGG-19+ResNet-50+VGG16	**91.35**	**92.65**

4.6. Additional Experiments

In this section, we study the importance of hyper-parameters and investigate the effect of changing components to provide additional insight into NLD.

Table 10 presents the following four experiments on UC Merced Land-use: (a) NLD with the weight factors $\alpha = 10$ and $\beta = 2$. (b) NLD with the weight factors $\alpha = 2$ and $\beta = 10$. (c) Using two same networks as student and teacher, respectively. (d) For the noisy teacher network, CE loss is replaced by SCE loss.

Table 10. Classification accuracy (%) on the UC Merced Land-use test set after changing each module from our model.

Network Types	Loss		None	Symm		Asym	
	α	β	$D_c : D_n$	$D_c : D_n$		$D_c : D_n$	
			10 : 0	6 : 4	4 : 6	4 : 6	2 : 8
ResNet-50+ResNet-34	$10CE$	$2CE$	99.08 ± 0.40	97.43 ± 0.63	**95.86 ± 0.29**	97.57 ± 0.49	97.38 ± 0.56
ResNet-50+ResNet-34	$2CE$	$10CE$	99.10 ± 0.31	95.33 ± 0.52	92.00 ± 0.80	**98.71 ± 0.27**	97.76 ± 0.71
ResNet-34+ResNet-34	$10CE$	$2CE$	98.00 ± 0.44	**97.48 ± 0.68**	95.29 ± 0.66	97.05 ± 1.05	97.33 ± 0.41
ResNet-50+ResNet-34	$10CE$	$2SCE$	**99.14 ± 0.12**	95.62 ± 0.75	93.28 ± 0.82	98.43 ± 0.24	**98.00 ± 0.44**

Hyper-parameters: From Table 10, hyper-parameters settings have a significant effect on the performance of NLD. As α decreases and β increases, the student network learns more information from noise distillation. Since the information in symmetric noise labels is limited, a larger β cannot make the teacher network to distill more knowledge. In such cases, the network performance can be degraded by incorrect guidance. Similarly, asymmetric noise labels have more correct information. So a larger β can enhance the teacher's ability to distill the right instruction to the student. In the absence of noisy labels, the effect of factors is not significant. This result thus suggests that appropriate factors are needed to select based on the quality of the noise labels in practice.

Distillation with the same network: As shown in Table 10, we perform experiments for ResNet-34 as a teacher and a student. In general, the first thing to notice is that the teacher network with a smaller capacity can also benefit the student network. However, for noise-free scenarios, it cannot take effect because the teacher and student have the same input and architecture and it is difficult to get extra knowledge. Moreover, a larger standard deviation for most results implies worse robustness. Therefore, a large teacher network is still a better option. In some low-cost scenarios, it is also possible to choose a small teacher network.

Training teacher with different loss: SCE can supervise the network to learn more information in the noisy labels (i.e., more errors in symmetric noise or more correctness in asymmetric noise). For fully clean data, there is little additional benefit from SCE. Such a property produces the results in Table 10. Therefore, for most real applications, SCE should be used instead of CE for the teacher to achieve a better performance of NLD.

5. Conclusions

This work proposes an efficient framework named NLD to address the noisy label problem for remote sensing image scene classification. NLD can distill the knowledge from different types of noise to improve performance of networks. Teacher networks can avoid overfitting into the noise through consistent decisions with student networks. The decision network is introduced to replace KL divergence. It is different from previous methods for distillation. The proposed NLD framework is end-to-end and does not require a pre-training process besides ImageNet. Thus, NLD is more practical and easier to deploy.

NLD can fully leverage the information contained in the noisy labels to improve the performance of network trained on the clean labels. Experiments are conducted on UC Merced Land-use, NWPU-RESISC45 and AID with different noise types. NLD improves over the baseline and direct fine-tuning. It can also be easily extended to pseudo-labeling. NLD performs significantly better than SSGA-E and other methods. For completely clean datasets, NLD can also improve accuracy as a model distillation-like process.

Future work will explore real-world noise datasets. More data with noisy labels can be collected from search engines and google earth, etc. Furthermore, mixing multiple existing public datasets as a clean dataset is also a worthwhile experiment. Our goal is to apply NLD to real scenarios.

Author Contributions: Conceptualization, R.Z., Z.C. and T.L. ; methodology, R.Z., Z.C. and T.L.; software, R.Z. and Z.C.; validation, R.Z., Z.C. and T.L.; formal analysis, R.Z. and F.S.; investigation, R.Z. and S.Z.; resources, T.L.; data curation, R.Z.; writing—original draft preparation, R.Z.; writing—review and editing, Z.C. and G.Z.; visualization, R.Z. and Q.Z.; supervision, T.L.; project administration, T.L.; funding acquisition, T.L. All authors have read and agreed to the published version of the manuscript.

Funding: This work was supported by Youth Innovation Promotion Association, Chinese Academy of Sciences (Grant No. 2016336).

Acknowledgments: The authors thank the editor and anonymous reviewers for their helpful comments and valuable suggestions.

Conflicts of Interest: The authors declare no conflict of interest.

Abbreviations

The following abbreviations are used in this manuscript:

MDPI	Multidisciplinary Digital Publishing Institute
DOAJ	Directory of open access journals
TLA	Three letter acronym
LD	linear dichroism

References

1. Chaib, S.; Liu, H.; Gu, Y.; Yao, H. Deep feature fusion for VHR remote sensing scene classification. *IEEE Trans. Geosci. Remote Sens.* **2017**, *55*, 4775–4784. [CrossRef]
2. Nogueira, K.; Penatti, O.A.; Dos Santos, J.A. Towards better exploiting convolutional neural networks for remote sensing scene classification. *Pattern Recognit.* **2017**, *61*, 539–556. [CrossRef]
3. Deng, J.; Dong, W.; Socher, R.; Li, L.; Li, K.; Li, F. ImageNet: A large-scale hierarchical image database. In Proceedings of the 2009 IEEE Computer Society Conference on Computer Vision and Pattern Recognition (CVPR), Miami, FL, USA, 20–25 June 2009; IEEE Computer Society: Washington, DC, USA, 2009; pp. 248–255.
4. Algan, G.; Ulusoy, I. Image Classification with Deep Learning in the Presence of Noisy Labels: A Survey. *arXiv* **2019**, arXiv:1912.05170.
5. Kuznetsova, A.; Rom, H.; Alldrin, N.; Uijlings, J.; Krasin, I.; Pont-Tuset, J.; Kamali, S.; Popov, S.; Malloci, M.; Duerig, T.; et al. The open images dataset v4: Unified image classification, object detection, and visual relationship detection at scale. *arXiv* **2018** arXiv:1811.00982.

6. Han, W.; Feng, R.; Wang, L.; Cheng, Y. A semi-supervised generative framework with deep learning features for high-resolution remote sensing image scene classification. *ISPRS-J. Photogramm. Remote Sens.* **2018**, *145*, 23–43. [CrossRef]
7. Li, W.; Wang, L.; Li, W.; Agustsson, E.; Van Gool, L. Webvision database: Visual learning and understanding from web data. *arXiv* **2017**, arXiv:1708.02862.
8. Xiao, T.; Xia, T.; Yang, Y.; Huang, C.; Wang, X. Learning from massive noisy labeled data for image classification. In Proceedings of the IEEE Conference on Computer Vision and Pattern Recognition (CVPR), Boston, MA, USA, 7–12 June 2015; IEEE Computer Society: Washington, DC, USA, 2015; pp. 2691–2699.
9. Lee, K.; He, X.; Zhang, L.; Yang, L. CleanNet: Transfer Learning for Scalable Image Classifier Training with Label Noise. In Proceedings of the 2018 IEEE Conference on Computer Vision and Pattern Recognition (CVPR), Salt Lake City, UT, USA, 18–22 June 2018; IEEE Computer Society: Washington, DC, USA, 2018; pp. 5447–5456.
10. Jiang, J.; Ma, J.; Wang, Z.; Chen, C.; Liu, X. Hyperspectral image classification in the presence of noisy labels. *IEEE Trans. Geosci. Remote Sens.* **2018**, *57*, 851–865. [CrossRef]
11. Tu, B.; Zhang, X.; Kang, X.; Wang, J.; Benediktsson, J.A. Spatial density peak clustering for hyperspectral image classification with noisy labels. *IEEE Trans. Geosci. Remote Sens.* **2019**, *57*, 5085–5097. [CrossRef]
12. Damodaran, B.B.; Flamary, R.; Seguy, V.; Courty, N. An Entropic Optimal Transport loss for learning deep neural networks under label noise in remote sensing images. *Comput. Vis. Image Underst.* **2020**, *191*, 102863. [CrossRef]
13. Wang, Y.; Ma, X.; Chen, Z.; Luo, Y.; Yi, J.; Bailey, J. Symmetric Cross Entropy for Robust Learning with Noisy Labels. In Proceedings of the 2019 IEEE/CVF International Conference on Computer Vision (ICCV), Seoul, Korea, 27 October–2 November 2019; IEEE: Piscataway, NJ, USA, 2019; pp. 322–330.
14. Northcutt, C.G.; Jiang, L.; Chuang, I.L. Confident Learning: Estimating Uncertainty in Dataset Labels. *arXiv* **2019**, arXiv:1911.00068.
15. Kim, Y.; Yim, J.; Yun, J.; Kim, J. NLNL: Negative Learning for Noisy Labels. In Proceedings of the 2019 IEEE/CVF International Conference on Computer Vision (ICCV), Seoul, Korea, 27 October–2 November 2019; IEEE: Piscataway, NJ, USA, 2019; pp. 101–110.
16. Veit, A.; Alldrin, N.; Chechik, G.; Krasin, I.; Gupta, A.; Belongie, S.J. Learning from Noisy Large-Scale Datasets with Minimal Supervision. In Proceedings of the 2017 IEEE Conference on Computer Vision and Pattern Recognition (CVPR), Honolulu, HI, USA, 21–26 July 2017; IEEE Computer Society: Washington, DC, USA, 2017; pp. 6575–6583.
17. Li, Y.; Yang, J.; Song, Y.; Cao, L.; Luo, J.; Li, L. Learning from Noisy Labels with Distillation. In Proceedings of the IEEE International Conference on Computer Vision (ICCV),Venice, Italy, 22–29 October 2017; IEEE Computer Society: Washington, DC, USA, 2017; pp. 1928–1936.
18. Hu, M.; Han, H.; Shan, S.; Chen, X. Weakly Supervised Image Classification Through Noise Regularization. In Proceedings of the IEEE Conference on Computer Vision and Pattern Recognition (CVPR), Long Beach, CA, USA, 16–20 June 2019; Computer Vision Foundation/IEEE: Piscataway, NJ, USA, 2019; pp. 11517–11525.
19. Zhang, Y.; Xiang, T.; Hospedales, T.M.; Lu, H. Deep Mutual Learning. In Proceedings of the 2018 IEEE Conference on Computer Vision and Pattern Recognition (CVPR), Salt Lake City, UT, USA, 18–22 June 2018; IEEE Computer Society: Washington, DC, USA, 2018; pp. 4320–4328.
20. Zagoruyko, S.; Komodakis, N. Learning to compare image patches via convolutional neural networks. In Proceedings of the IEEE Conference on Computer Vision and Pattern Recognition (CVPR), Boston, MA, USA, 7–12 June 2015; IEEE Computer Society: Washington, DC, USA, 2015; pp. 4353–4361.
21. Yang, Y.; Newsam, S.D. Bag-of-visual-words and spatial extensions for land-use classification. In Proceedings of the 18th ACM SIGSPATIAL International Symposium on Advances in Geographic Information Systems (ACM-GIS), San Jose, CA, USA, 3–5 November 2010; Agrawal, D., Zhang, P., Abbadi, A.E., Mokbel, M.F., Eds.; ACM: New York, NY, USA, 2010; pp. 270–279.
22. Cheng, G.; Han, J.; Lu, X. Remote sensing image scene classification: Benchmark and state of the art. *Proc. IEEE* **2017**, *105*, 1865–1883. [CrossRef]
23. Xia, G.S.; Hu, J.; Hu, F.; Shi, B.; Bai, X.; Zhong, Y.; Zhang, L.; Lu, X. AID: A benchmark data set for performance evaluation of aerial scene classification. *IEEE Trans. Geosci. Remote Sens.* **2017**, *55*, 3965–3981. [CrossRef]

24. He, K.; Zhang, X.; Ren, S.; Sun, J. Deep Residual Learning for Image Recognition. In Proceedings of the 2016 IEEE Conference on Computer Vision and Pattern Recognition (CVPR), Las Vegas, NV, USA, 27–30 June 2016; IEEE Computer Society: Washington, DC, USA, 2016; pp. 770–778.
25. Simonyan, K.; Zisserman, A. Very deep convolutional networks for large-scale image recognition. *arXiv* **2014**, arXiv:1409.1556.
26. Zhang, W.; Tang, P.; Zhao, L. Remote sensing image scene classification using CNN-CapsNet. *Remote Sens.* **2019**, *11*, 494. [CrossRef]
27. Li, J.; Lin, D.; Wang, Y.; Xu, G.; Zhang, Y.; Ding, C.; Zhou, Y. Deep Discriminative Representation Learning with Attention Map for Scene Classification. *Remote Sens.* **2020**, *12*, 1366. [CrossRef]
28. Inoue, N.; Simo-Serra, E.; Yamasaki, T.; Ishikawa, H. Multi-label Fashion Image Classification with Minimal Human Supervision. In Proceedings of the 2017 IEEE International Conference on Computer Vision Workshops (ICCV Workshops), Venice, Italy, 22–29 October 2017; IEEE Computer Society: Washington, DC, USA, 2017; pp. 2261–2267.
29. Sohn, K.; Berthelot, D.; Li, C.L.; Zhang, Z.; Carlini, N.; Cubuk, E.D.; Kurakin, A.; Zhang, H.; Raffel, C. Fixmatch: Simplifying semi-supervised learning with consistency and confidence. *arXiv* **2020**, arXiv:2001.07685.
30. Li, Q.; Peng, X.; Cao, L.; Du, W.; Xing, H.; Qiao, Y. Product Image Recognition with Guidance Learning and Noisy Supervision. *arXiv* **2019**, arXiv:1907.11384.
31. Hinton, G.; Vinyals, O.; Dean, J. Distilling the knowledge in a neural network. *arXiv* **2015**, arXiv:1503.02531.
32. Ioffe, S.; Szegedy, C. Batch normalization: Accelerating deep network training by reducing internal covariate shift. *arXiv* **2015**, arXiv:1502.03167.
33. Srivastava, N.; Hinton, G.; Krizhevsky, A.; Sutskever, I.; Salakhutdinov, R. Dropout: A simple way to prevent neural networks from overfitting. *J. Mach. Learn. Res.* **2014**, *15*, 1929–1958.
34. Chatfield, K.; Simonyan, K.; Vedaldi, A.; Zisserman, A. Return of the devil in the details: Delving deep into convolutional nets. *arXiv* **2014**, arXiv:1405.3531.
35. Paszke, A.; Gross, S.; Massa, F.; Lerer, A.; Bradbury, J.; Chanan, G.; Killeen, T.; Lin, Z.; Gimelshein, N.; Antiga, L.; et al. PyTorch: An Imperative Style, High-Performance Deep Learning Library. In Proceedings of the Advances in Neural Information Processing Systems 32: Annual Conference on Neural Information Processing Systems 2019 (NeurIPS), Vancouver, BC, Canada, 8–14 December 2019; Wallach, H.M., Larochelle, H., Beygelzimer, A., d'Alché-Buc, F., Fox, E.B., Garnett, R., Eds.; Curran Associates, Inc.: Red Hook, NY, USA, 2019; pp. 8024–8035.

MDPI

Technical Note

Feature-Level Fusion of Polarized SAR and Optical Images Based on Random Forest and Conditional Random Fields

Yingying Kong [1,*], Biyuan Yan [1], Yanjuan Liu [1], Henry Leung [2] and Xiangyang Peng [3]

1 College of Electronic and Information Engineering, Nanjing University of Aeronautics and Astronautics, Nanjing 210016, China; yanbiyuan@nuaa.edu.cn (B.Y.); liuyanjuan@nuaa.edu.cn (Y.L.)
2 Department of Electrical and Computer Engineering, University of Calgary, Calgary, AB T2P 2M5, Canada; Leungh@ucalgary.ca
3 Nanjing Research Institute of Electronics Engineering, Nanjing 210007, China; wwukt@163.com
* Correspondence: yayako_zy@nuaa.edu.cn; Tel.: +86-1855-140-8717

Abstract: In terms of land cover classification, optical images have been proven to have good classification performance. Synthetic Aperture Radar (SAR) has the characteristics of working all-time and all-weather. It has more significant advantages over optical images for the recognition of some scenes, such as water bodies. One of the current challenges is how to fuse the benefits of both to obtain more powerful classification capabilities. This study proposes a classification model based on random forest with the conditional random fields (CRF) for feature-level fusion classification using features extracted from polarized SAR and optical images. In this paper, feature importance is introduced as a weight in the pairwise potential function of the CRF to improve the correction rate of misclassified points. The results show that the dataset combining the two provides significant improvements in feature identification when compared to the dataset using optical or polarized SAR image features alone. Among the four classification models used, the random forest-importance_conditional random fields (RF-Im_CRF) model developed in this paper obtained the best overall accuracy (OA) and Kappa coefficient, validating the effectiveness of the method.

Keywords: polarized SAR; optical image; random forest; conditional random fields; feature-level fusion

Citation: Kong, Y.; Yan, B.; Liu, Y.; Leung, H.; Peng, X. Feature-Level Fusion of Polarized SAR and Optical Images Based on Random Forest and Conditional Random Fields. *Remote Sens.* **2021**, *13*, 1323. https://doi.org/10.3390/rs13071323

Academic Editor: Monidipa Das

Received: 1 March 2021
Accepted: 24 March 2021
Published: 30 March 2021

1. Introduction

The impact of urban development on the Earth's environment is enormous, leaving an ever-changing imprint on its surface. This situation calls for a compulsory requirement to map the land cover and review land-use patterns of our dynamic eco-system time [1]. Polarized Synthetic Aperture Radar (SAR) and optical image have gained many applications in land cover classifications [2–5]. Since the two have entirely different physical properties, this makes them have distinct advantages in classification. For example, the optical images are susceptible to differences in the vegetation spectrum and are, therefore, often used to detect pest and disease problems [6]. SAR images offer high accuracy and purity in detecting water areas, but extracting sharp edges is still a challenge [7]. Therefore, how to fully utilize the advantages of both is one of the major topics currently faced.

Data fusion is a way to take full advantage of multiple sources of data. The data fusion stages (pixel-level, feature-level, and decision-level) determine the data fusion techniques [8]. Feature-level fusion consists of two critical processes: image feature extraction and feature merging. In this regard, Aswatha et al. [1] used multimodal information from multispectral images and polarized SAR data to classify land cover into seven classes in an unsupervised manner. Su [9] extracted the backward scattering features and grey-level co-occurrence matrix (GLCM) features obtained from the Pauli decomposition and H/A/alpha decomposition of polarized SAR images, the spectral features, and GLCM features of multispectral images, and used a support vector machine (SVM) for classification.

This fusion method effectively improves the classification accuracy and the pepper noise is reduced.

Land cover classification is one of the critical applications of remote sensing images. The traditional land cover classification method is divided into two steps: feature extraction and classifier training [10].

The feature extraction for optical images is based on spectral and textural features. A textural feature is a comprehensive reflection of the image greyscale statistical information, spatial distribution information, and structural information. Commonly used textural feature classification algorithms include a local binary pattern (LBP) [11], GLCM [12], etc. Polarized SAR feature extraction is based on polarized target decomposition, which aims to decode the scattering mechanism of the feature under a reasonable physical constraint model [13], such as Freeman-Durden decomposition [14], Yamaguchi decomposition [15], etc.

Machine learning has achieved considerable progress in classification and regression tasks. Commonly used machine learning is SVM, decision tree, random forest, etc. In the current research, SVM has been used extensively. For example, Attarchi [16] used SVM to classify polarized SAR data and its GLCM features for the detection of impervious surfaces. While SVM classifies samples by finding hyperplanes, decision trees classify samples by selecting the optimal components and dividing the subset into the corresponding leaf nodes based on the features. Phartiyal et al. [17] used an evolutionary genetic algorithm to optimize the empirical model to maximize the classification performance. They constructed a decision tree based on the best class boundary and obtained satisfactory classification results. Random forest is an ensemble learning model based on decision trees, which obtains the final results by combining and analysing multiple decision trees [18]. Du et al. [19] extracted the polarization and texture features of the fully polarized SAR images for random forest and rotation forest classifiers. The experiment finally verified that random forest is better than Wishart and SVM classifiers, and it is less accurate than rotation forest but faster.

In image processing, conditional random fields (CRF) have unique advantages in expressing the spatial context and the posterior probability modelling [20]. Zhong et al. [21] proposed the spatial-spectral-emissivity land-cover classification based on the conditional random fields (SSECRF) algorithm, which integrates the spatial-spectral feature set and emissivity by constructing the SSECRF energy function to obtain better classification results. CRF allows for the processing of target classes in conjunction with neighbourhood information, effectively improving the image purity of the classification results, which is missing from machine learning.

This article proposes an RF-Im_CRF classification model to improve the accuracy of the random forest classifier in feature-level fusion classification. The model first extracts the spectral and GLCM features of optical images, the Freeman decomposition, and Polarization Signature Correlation Feature (PSCF) features of polarized SAR. Then, the model assembled them into a random forest training dataset. Afterward, the random forest classifier results are input into the Im_CRF model, which uses the feature importance from the random forest as the weight information in the pairwise potential function to improve feature classification accuracy.

2. Materials

2.1. Study Site

The location selected for this study is in Nanjing and its surrounding area, which is located in Jiangsu Province in Eastern China. Figure 1 shows the optical and polarized SAR false-colour images of the study area. The false-colour image is generated based on the Pauli decomposition. The images are 1500 × 1500 pixels in size, which include river, buildings, vegetation, and roads. The image resolution is 8 metres, so the total size of the study area is about 169 km^2. The architective area occupies the majority of the image, the vegetation area is relatively concentrated, and there is a small amount of vegetation within the building space. The cultivated area is concentrated in the northern part of the river. A clear colour difference can be observed in the optical image between the dense vegetation area and the cultivated area. The colour of the river part is not sufficiently uniform, which is similar to the farmland in some areas. In contrast, the river area of the SAR false-colour image is different from other regions. Therefore, it can be seen that polarized SAR has apparent advantages in identifying river categories.

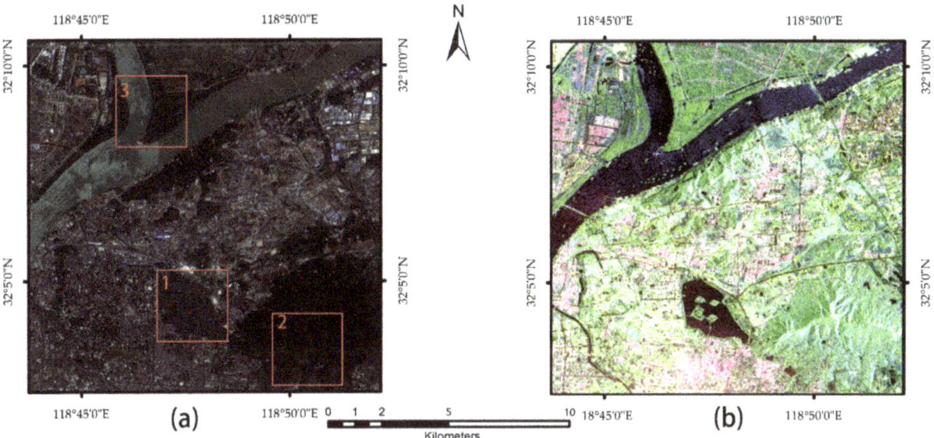

Figure 1. Study area. (**a**) The optical image. (**b**) The polarized SAR false-colour image.

The dataset used for research is the polarized SAR data collected by the RADARSAT-2 satellite, which has four polarization states: HH, VV, HV, and VH. This data was acquired on 19 April 2011 at a resolution of 8 m. The optical image resolution is 5 m, and the acquisition time is April 2017. Due to the relatively low resolution and the fact that the acquisition time falls within the same month, the variation in ground objects is within manageable limits. In the ENVI software, the optical image was down-sampled to a resolution of 8 m, and the polarized SAR image has undergone preprocessing such as multi-looking and noise reduction. The two images were calibrated in the same geographic coordinate system.

2.2. Sampling Point Selection

The sampling point coordinates in the experiment were taken with the optical image as a reference. Overall, five land cover categories were considered, namely Water, Building, High vegetation, Low vegetation, and Road. The high vegetation is dominated by tall forests and the low vegetation is dominated by agricultural land. Since the image resolution is 8 m, this prevents some narrow roads from being clearly represented, especially for SAR images. This paper, therefore, chose to sample roads with larger width, such as motorways and arterial roads. Because of the massive amount of source image data, it is not easy to classify the entire image finely. Therefore, the training samples chosen for this experiment are 100 per class, and the test samples are 150 for each category, as shown in Table 1. The totals of training samples and test samples are 500 and 750, respectively, with no duplicate points.

Table 1. Sample label type and quantity.

Label Category	Train Number	Test Number
Water	100	150
High vegetation	100	150
Building	100	150
Low vegetation	100	150
Road	100	150
Total	500	750

3. Characteristic Data Acquisition

3.1. Polarization Feature Extraction

For the extraction of polarized SAR image features, this experiment selected two polarization feature extraction methods known as the Freeman-Durden decomposition and the PSCF.

3.1.1. Freeman-Durden Decomposition

The Freeman-Durden polarization decomposition method is based on the fundamental principle of radar scattering, which decomposes the SAR cross-covariance matrix into canopy scattering (or volume scattering), odd bounce scattering (or surface scattering), and double-bounce scattering (or dihedral scattering). The detailed description of the modelling process for the composite scattering model can be found in Reference [22]. This model can acquire the characteristic parameters related to the three scattering mechanisms and the corresponding weight coefficients.

The power corresponding to the three scattering mechanisms are Ps, Pd, and Pv, where Ps corresponds to the power of surface scattering, Pd represents the power of dihedral scattering, and Pv represents the power of volume scattering. Then, the Freeman feature vector of the target points can be established.

$$X_{Freeman} = [x_i^{Pd}, x_i^{Ps}, x_i^{Pv}]^T \tag{1}$$

3.1.2. Polarization Signature Correlation Feature (PSCF)

Radar polarization signatures (PSs) can effectively characterize the scattering behaviour of the research object, so it has the potential to distinguish the types of ground objects. This feature is usually a three-dimensional representation of the backscattering behaviour of a target or land cover. In the expression of PSs, the x-axis and y-axis represent the ellipse angle and azimuth angle, respectively, and the z-axis represents the received backscattering power coefficient. The value range of the azimuth angle (ψ) is -90 to 90 degrees, and the value range of the ellipse angle (χ) is -45 to 45 degrees. The following formula gives the PSs.

$$\sigma(\chi_i\psi_i\chi_j\psi_j) = \frac{4\pi}{k^2}\begin{pmatrix} 1 \\ \cos 2\chi_i \cos 2\psi_i \\ \cos 2\chi_i \sin 2\psi_i \\ \sin 2\chi_i \end{pmatrix}(K)\begin{pmatrix} 1 \\ \cos 2\chi_j \cos 2\psi_j \\ \cos 2\chi_j \sin 2\psi_j \\ \sin 2\chi_j \end{pmatrix} \tag{2}$$

Among them, σ represents the backscattering coefficient or received power, the subscripts i and j mean the transmitting and receiving units, respectively, and K is the Kennaugh matrix [23]. k is the wave number of the illuminating wave.

The co-polarized signatures are obtained by transmitting and receiving combination $\psi_i = \psi_j, \chi_i = \chi_j$, and the cross-polarized signatures are obtained by $\psi_i = 90 + \psi_j, \chi_i = -\chi_j$. The ellipse angle defines the polarization behaviour (linear polarization, circular polarization, or elliptical polarization), and the azimuth angle defines the polarization states, that is, horizontal or vertical polarization [24]. In the current research, the characteristics of co-polarized and cross-polarized signatures have been fully considered and utilized.

Since surface objects generally exhibit a complex scattering response, the polarization signatures of standard targets must be used as a reference for classification. Therefore, PSs have been calculated for flat plate (FP), horizontal dipole (HD), vertical dipole (VD), and a dihedral angle (Di) in the standard targets. The formulae for the generation of the standard target PSs are given in Reference [25].

Therefore, the PSCF uses the radar polarization signatures of the four standard scatterers (FP, HD, HD, and VD) as a reference to calculate the relevance between the polarization characteristics of the target points and the above four standard targets. This can be a reference to distinguish between different categories. The correlation coefficient formula is as follows.

$$CC = \frac{S_{xy}}{S_x S_y} \tag{3}$$

where x and y are the polarized characteristics of the target points and the standard targets, respectively. S_x is the standard deviation of x, S_y is the standard deviation of y, and S_{xy} is the covariance between x and y. CC is the correlation coefficient between x and y.

This paper refers to Reference [17] to obtain the PSCF solution and establish the feature correlation coefficients between a single target and four standard targets, which are Corr_co_Di, Corr_co_FP, Corr_co_HD, Corr_co_VD, Corr_cross_Di, Corr_ cross _FP, Corr_ cross_HD, and Corr_ cross _VD. Among them, the co is for the co-polarization while the cross is for cross-polarization. Thus, the PSCF feature vector of the target point is established as:

$$\begin{aligned} X_{PSCF} = [&x_i^{corr\_co\_Di}, x_i^{corr\_co\_FP}, x_i^{corr\_co\_HD}, x_i^{corr\_co\_VD}, \\ &x_i^{corr\_cross\_Di}, x_i^{corr\_cross\_FP}, x_i^{corr\_cross\_HD}, x_i^{corr\_cross\_VD}]^T \end{aligned} \tag{4}$$

3.2. *Optical Image Feature Extraction*

3.2.1. Spectral Information Extraction

Compared with multispectral images, the optical image does not have rich spectral information, but it is also sufficient to identify information with significant spectral differences. This optical image can be divided into three bands: red, green, and blue, so the spectral feature information is shown as follows.

$$X_{Spectral} = [x_i^r, x_i^g, x_i^b]^T \tag{5}$$

3.2.2. Grey-Level Co-Occurrence Matrix (GLCM)

The textural feature is a visual feature that does not depend on brightness and colour, reflecting similar information of adjacent pixels in the image. It reflects the internal characteristics shared by the surface of the object. It contains essential information about the surface structure of the object and the relationship to its neighbours.

GLCM is a commonly used method for extracting texture information with good discrimination ability. Its principle is to convert the specified spatial relationship in the image into texture information based on the greyscale value. The texture features obtained by GLCM are helpful to distinguish objects with similar spectral characteristics.

In this paper, three features are chosen to describe the spatial relationships of images: contrast, dissimilarity, and energy. Contrast and dissimilarity can measure the local variation and reflect the sharpness of the image and the depth of the texture. The energy is the sum of the squares of element values of the GLCM, demonstrating the uniformity of the image greyscale distribution and the texture thickness. The GLCM feature information is expressed as follows.

$$X_{GLCM} = [x_i^{contrast}, x_i^{dissimiliraty}, x_i^{energy}]^T \tag{6}$$

4. Random Forest-Importance_Conditional Random Forest (RF-Im_CRF) Model

Figure 2 is the flowchart of applying the RF-Im_CRF model to the feature-level fusion of polarized SAR and optical images. After extracting the features of the two images, the random forest is first used for classification. Then, the classification results and feature importance of the random forest are combined with the CRF. The classification results are taken as the unary potential function and the feature importance is taken as the weight of the pairwise potential function to improve the classification accuracy.

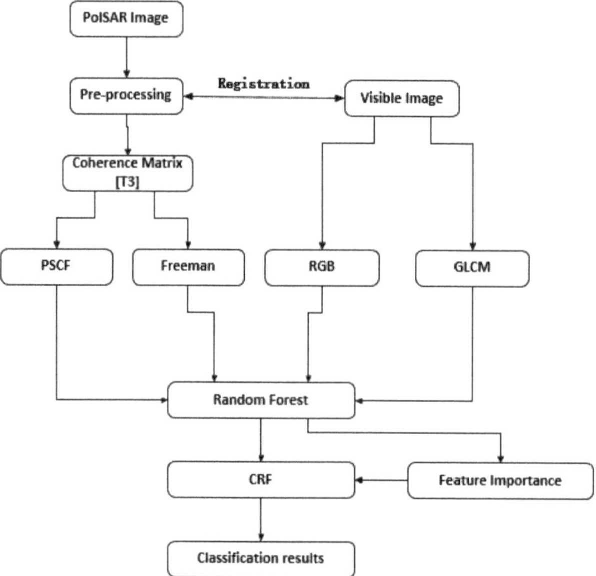

Figure 2. RF-Im_CRF model flowchart.

4.1. Random Forest

Random forests construct mutually independent decision trees in which each generates a training set by bootstrap resampling. M rounds were randomly selected from the original training set with N samples to obtain M training sets. Some samples may be chosen multiple times under self-service resampling, while some samples may not be drawn. Then M decision trees are developed according to these training sets. In the decision-making stage, the classification results are obtained by taking the mode, or the regression results, by taking the average value. The random forest can process large data sets with high efficiency and precision, filter explanatory variables by itself, and get the mutual influence and importance ranking of variables.

The Gini index, or Gini impurity, indicates the probability that a randomly selected sample in the sample set will be misclassified. At each node in the binary tree T of the random forest, the optimal segmentation is sought according to the Gini index $i(\tau)$, which divides the sub-node data set. Random forest follows the principle of Gini gain maximization when selecting features for nodes [26]. Let p_k be the probability of node τ being divided into child nodes $\tau_k, k = 1, 2$. Then the Gini index is:

$$i(\tau) = \sum_{k=1}^{2} p_k(1 - p_k) = 1 - \sum_{k=1}^{2} p_k^2 \tag{7}$$

The Gini gain Δi generated by splitting the sample through a certain threshold and sending it to two child nodes τ_1 and τ_2, which is defined as:

$$\Delta i(\tau) = i(\tau) - p_1 i(\tau_1) - p_2 i(\tau_2) \tag{8}$$

Since the decision tree selects features that can maximize the Gini gain of the node when generating nodes, the feature importance can be reflected by the sample division of the nodes. However, random forest introduces the double randomness of data samples and input features during a training process, which may cause important features with high discrimination being used to divide nodes less frequently than features with low discrimination. Therefore, the importance of features cannot be measured simply by the number of times used as segmentation attributes [27,28].

4.2. Conditional Random Fields

The CRF model simulates the local neighbourhood interaction between random variables in the unified probability framework. Given the observed image data, the model directly models the posterior probability of the label as a Gibbs distribution.

The general form of the CRF model is:

$$P(Y|X) = \frac{1}{Z(X)} exp\left\{ - \left[\sum_{i \in V} \Phi\%_i(y_i, x_i, w) + \beta \sum_{(i,j) \in E} \Phi\%_{ij}(y_i, y_j, x_i, x_j, v) \right] \right\} \tag{9}$$

Among them, V is for the set of data points and E is for the set of point neighbours. x_i, y_i represents the observation variable of the i-th point in the data and its class label variable, respectively. X is the sequence of observations, $X = [x_1, \ldots, x_i, \ldots, x_N]$. Y is the sequence of tags corresponding to X, $Y = (y_1, \ldots, y_i, \ldots, y_C)$, where C is the number of categories. $P(Y|X)$ is the probability of the label sequence Y under the given observation sequence X. $Z(X)$ is the normalization constant, $Z(X) = \sum_Y exp\left\{ - \sum_{c \in C} \Phi\%_c(y_c, x) \right\}$; $\Phi\%_i(\cdot)$ is the unary potential function, which represents the probability of the observed variable x_i taking the label y_i. $\Phi_{ij}(\cdot)$ is the pairwise potential function, which means the correlation between the variable x_i and its neighbouring variables x_j and the correlation between the labels. w, v, respectively, represents the parameters of the correlation potential function and the interaction potential function. β is to adjust the weight of the two potential function terms, which determines the degree of influence of the pairwise function on the unray potential function. In this article, to simplify the implementation of CRF, β is set to a constant 1.

Then the corresponding Gibbs energy is defined as:

$$\begin{aligned} E(Y|X) &= -\log P(Y|X) - \log(Z(X)) = \sum_{c \in C} \Phi\%(y_c, x) \\ &= \sum_{i \in V} \Phi\%_i(y_i, x_i, w) + \beta \sum_{(i,j) \in E} \Phi\%_{ij}(y_i, y_j, x_i, x_j, v) \end{aligned} \tag{10}$$

According to the Bayesian Maximum Posterior (MAP) rule, image classification aims to find the label Y that maximizes the posterior probability $P(Y|X)$. Therefore, the CRF's MAP mark xMAP can be obtained by the following formula.

$$Y_{MAP} = arg \max_y P(Y|X) = arg \min_y E(Y|X) \tag{11}$$

It can be seen that finding the maximum value of the posterior probability $P(Y|X)$ is equivalent to finding the minimum value of the energy function $E(Y|X)$. Therefore, the optimization algorithm finds the most probable label by finding the minimum energy solution.

4.3. RF-Im_CRF Model

4.3.1. Establishment of Potential Functions

In this paper, the unary potential function $\Phi\%_i$ is defined based on the classification results of the random forest classifier. For variables x_i and its label y_i, when $y_i = k, \forall k \in K$ (K is the label set), then Equation (12) is:

$$P(y_i = k|x_i) = \frac{1}{M}\sum_{m=1}^{M}\delta[T_m(x_i, \theta_m) = k] \tag{12}$$

M is the total number of decision trees. θ_m is the independent and identically distributed parameter vector describing the m-th decision tree. Then, $P(y_i = k|x_i)$ represents the probability that the target is of class k.

The CRF unary potential function is defined as:

$$\Phi\%_i(y_i, x_i) = -logP(y_i|x_i) \tag{13}$$

Pairwise potential function $\Phi\%_{ij}(y_i, y_j, x_i, x_j, v)$, also called the smoothness term, encourages adjacent pixels of the image to use the same label. This article uses an improved contrast-sensitive Potts model that introduces the feature importance η_k to define the pairwise potential function.

$$\Phi\%_{ij}(y_i, y_j, x_i, x_j) = \begin{cases} 0 & if\ y_i = y_j \\ g_{ij}(S) & otherwise \end{cases} \tag{14}$$

$$g_{ij}(S) = dist(i,j)^{-1}exp\left(-\sum_{k=1}^{N}\eta_k\gamma_k \parallel X_i^k - X_j^k \parallel^2\right) \tag{15}$$

Among them, g_{ij} simulates the spatial interaction of adjacent pixels x_i and x_j, which is used to measure the feature difference between neighbours. $dist(i,j)$ is the Euclidean distance between adjacent pixels, X_i^k and X_j^k represent the feature vector between points i and j. k represents the category of the feature vector, namely, $k = 1,2,3,4$, which, respectively, represents the feature vector $X_{Freeman}, X_{PSCF}, X_{Spectral}, X_{GLCM}$. γ_k is set to be the mean square error of feature vectors between adjacent pixels in the image, denoted as $\gamma_k = \left(2\left\langle\left|\left|X_i^k - X_j^k\right|\right|^2\right\rangle\right)^{-1}$, which $\langle\cdot\rangle$ represents the mean value of the neighbourhood. The parameter η_k is the feature importance in the classification process, obtained by random forest.

4.3.2. Feature Importance

In this paper, the statistic Im_i is used as a feature importance measurement based on the Gini index, representing the average change in the Gini index of the i-th feature in the node division of all decision trees. The importance of feature x_i on node n is the change in the Gini index that the sample on the node τ is divided into child nodes τ_1 and τ_2 in which:

$$Im_{i,m,n} = i(\tau) - i(\tau_1) - i(\tau_2) \tag{16}$$

where $n = 1, \ldots, N$, which represents the node index in one decision tree, and $m = 1, \ldots, M$, which represents the decision tree index in the random forest. Therefore, the feature x_i has N nodes in the m-th decision tree as the attribute of node division. Then the feature importance x_i on this decision tree can be expressed as:

$$Im_{i,m} = \sum_{n=1}^{N} Im_{i,m,n} \tag{17}$$

The feature importance x_i in the entire random forest is:

$$Im_i = \frac{1}{M}\sum_{m=1}^{M} Im_{i,m} \tag{18}$$

The sum of the feature importance of each feature is 1.

For parameter η_k, Freeman decomposition, PSCF features, spectral features, and GLCM features are regarded as four various feature components. Then, taking spectral features as an example, the feature importance of this characteristic component is:

$$\eta_{Spectral} = Im_r + Im_g + Im_b \tag{19}$$

The four feature components extracted in this paper have different value ranges and number of elements. Since the normalization of features does not affect the random forest results, they are not normalized in feature extraction. However, in the CRF, this difference in the value range affects the pairwise potential function. Therefore, it needs to be divided into four parts to avoid the features with a small value range in which they do not work as well as they should. Since the importance of each feature is different, the higher the importance of the feature, the greater the influence on classification. Therefore, the parameters η_k can further strengthen the feature difference between neighbours and improve classification accuracy.

5. Experiment and Analysis

5.1. Multi-Source Data Comparative Classification Experiment

First, to verify the advantages of image fusion in image classification, this paper used the random forest to perform classification experiments on optical image data and polarized SAR data. The optical image data contains a feature vector consisting of spectral and GLCM information, and the polarized SAR data includes a feature vector consisting of Freeman and PSCF information. The number of decision trees in the random forest was set to 100. This value ensures that the results of the random forest will be optimal and fluctuate within a range of values. The experimental results are shown as follows.

For classification tasks, the classification results can intuitively and clearly reflect the disparity between different features or different classification methods, especially when the distinction is significant. Figure 3 shows the classification results obtained by adopting different feature vectors. It can be seen that the characteristics of the optical image can better distinguish the difference between high and low vegetation due to the apparent differences in spectra. However, the reliance on spectral features also makes many errors in the identification of waters. Since the water surface tends to be specularly reflective, the backscatter from the water surface is almost zero, resulting in high accuracy of SAR image classification in waters. At the same time, the working frequency band of RADARSAT-2 is C-band, which has certain penetrability, making it difficult to distinguish the characteristic difference between high and low vegetation, thus, presenting a mixed phenomenon of dark green and light green. This penetrability is also reflected in the ability of the polarized SAR data to detect folds in the hills and present similar features to buildings, leading to misinterpretations. Optical image features have certain advantages in terms of buildings, and it is difficult for both sides to get ideal results on the road.

The visual effect of the classification that combines polarized SAR and optical image features is significantly improved. The water area as well as high and low vegetation are well inherited. Simultaneously, compared with the former two, the salt and pepper noise in the construction area has been significantly reduced. The large area of misjudgment is also hard to see, and the display effect of the road is improved. This indicates that the characteristics of polarized SAR and optical images both play a specific role in classification. Due to the similarity of the narrow river sections to the backscattering of the road, this caused the SAR data to misinterpret at the river in the southwest region of the image. This situation is also shown in Figure 3c. This indicates that the features of the optical images are still difficult to correct for the high misclassification of SAR images in this particular scene.

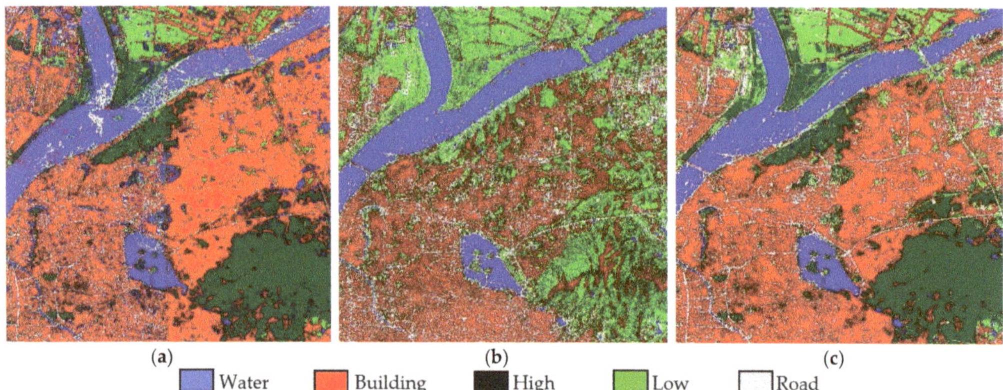

Figure 3. Multi-source data classification results. (**a**) Optical image classification result. (**b**) Polarized SAR classification result. (**c**) Optical + polarized SAR image classification result.

From the experimental results, it can be seen that the integrated polarized SAR and optical image fusion classification performance is significantly improved compared with the image classification performance of the single source. However, there are still many noise points, which affect the smoothness of the classification result. The RF-Im_CRF model proposed in this paper will improve the classification results aiming at this phenomenon.

5.2. Comparison of RF-Im_CRF Model Experiment Results

5.2.1. Analysis of Classified Image Results

To verify the effectiveness of the algorithm in this paper, the experimental data were classified using SVM based on Poly kernel function, RF, RF-CRF without feature importance as weights [21], and the RF-Im_CRF models, respectively. The experimental data is the feature vector composed of the four features in Chapter 3 of the article. The results are shown in Figure 4.

It can be seen that the SVM has the worst classification effect. SVM is an independent classifier, so it follows one rule when classifying. Random forests, on the other hand, rely on multiple mutually independent decision trees acting together, each with a different classification threshold. This means that the misclassification results of a single decision tree are corrected by the action of other decision trees. As a result, random forests give better results.

Compared with the random forest classifier, the RF-CRF model significantly improves image smoothness, since the CRF eliminates most salt and pepper noise. The differences between the RF-CRF and RF-Im_CRF models are difficult to see. Therefore, this paper extracted three scenes in the image for comparison to show the performance gap between the two models. The reference data are the optical image and the real classification results based on the optical image.

As shown in Figure 5, when compared with the RF-CRF model, the RF-Im_CRF model can further reduce the salt and pepper noise in the image, and the smoothness can be further improved. Since parking lots are set up around some large buildings, the classifier will be difficult to balance between roads and buildings. Some open places such as sports fields and squares as well as roads have more white blocks in area 1, which represent the road. Area 2 has lower category complexity and better homogeneity of vegetation, so there is less variation in the effects of classification. There are narrow roads in area 3, which were not sampled as samples during the sampling process, since it hardly distinguished with low contrast between neighbours in the SAR image. Therefore, it is misclassified as low vegetation in the classification result. The small white areas in the river are the ships sailing on the river in the SAR image. The RF-Im_CRF model is better than the RF-CRF model in identifying the riverbank portion on the left side, showing a relatively complete low vegetation zone.

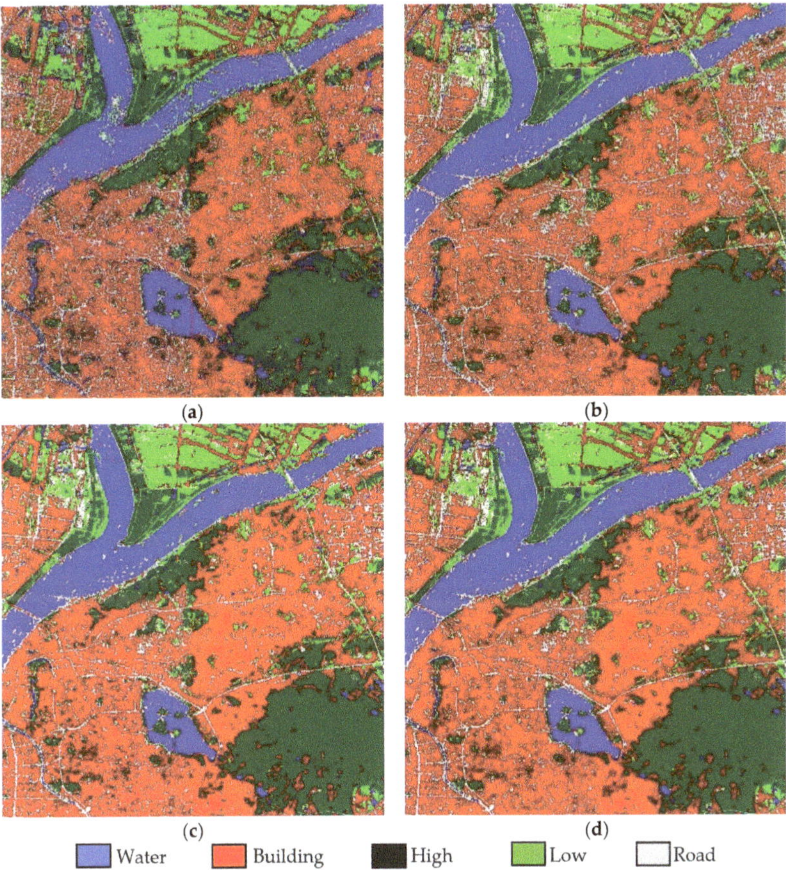

Figure 4. Classification results with different classifier (**a**) SVM classifier result, (**b**) RF classifier result, (**c**) RF-CRF model result, and (**d**) RF-Im_CRF model result.

The display of the classification results shows that, when compared with the RF-CRF model, the RF-Im_CRF further improves the classification accuracy, resulting in less noisy images and a further increase in purity. This is because the value range of various features is diverse. For example, the value range of the spectral feature is between 0–255, while the value range of PSCF is between -1 and 1. The feature difference is calculated in the unit of a feature component in CRF, which helps reduce the overall influence of features with a wide value range. Simultaneously, after adding feature importance as weights, the impact of features with high importance on feature differences between neighbours is enhanced. Therefore, the RF-Im_CRF model can classify ground objects more accurately.

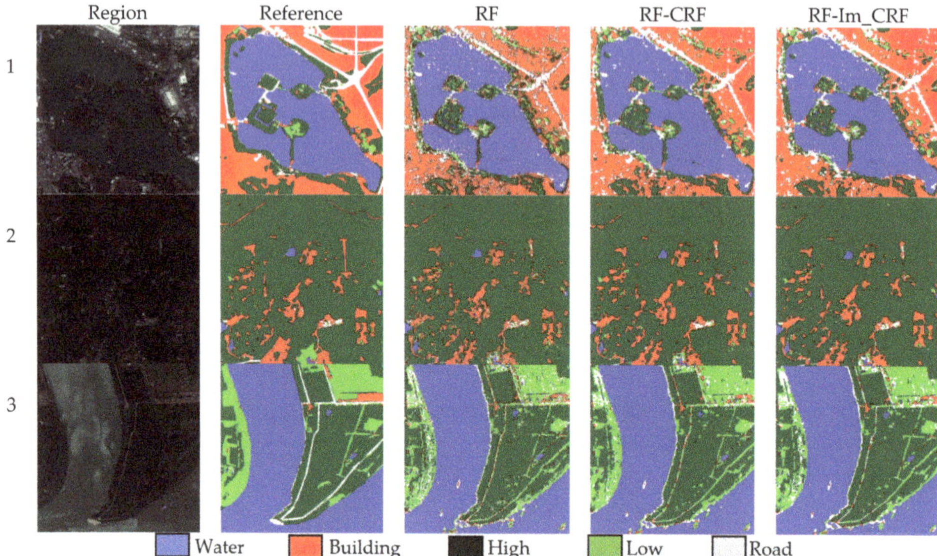

Figure 5. Sub-area classification results. The regions numbered 1, 2, and 3 are the corresponding subregions in Figure 1a.

5.2.2. Classification Data Analysis

This paper quantified the classification effectiveness of the classification model through Overall Accuracy (OA) and a Kappa coefficient, and analysed various classification cases using precision and recall.

When the training set is the same, the SVM produce the same results in multiple experiments. In contrast, the random forest has a certain degree of randomness. Even though the training set is the same, the results obtained during each training set are different. Therefore, we used the same dataset for ten consecutive tests on the random forest model to get the average of the results. In each experiment, the RF, RF-CRF, and RF-Im_CRF models use the same RF model results, which are only different in the subsequent processing. The RF model was built on Scikit-learn package using Python [29]. In each experiment, this paper extracted the feature importance and the probability of each class of all points. At the end, the evaluation index, such as OA and Kappa coefficients, were obtained for each model based on classification results.

The OA, Kappa values, and their 95% confidence interval are shown in Table 2.

Table 2. The average of OA and Kappa.

	SVM	RF	RF-CRF	RF-Im_CRF
OA 95% confidence interval	79%	88.0% [85.88%,90.4%]	91.6% [90.22%,93.02%]	94.0% [93.52%,94.54%]
Kappa 95% confidence interval	0.74	0.85 [0.834,0.866]	0.89 [0.879,0.905]	0.91 [0.902,0.918]

With the same test data and constant parameters, the results of the SVM are always consistent and, therefore, there are no confidence intervals. In terms of a quantitative data comparison, the RF-Im_CRF model proposed in this paper has the best classification accuracy with an average OA of 94.0%, and the 95% confidence interval is [93.52%,94.54%]. The Kappa coefficient is 0.91 with the 95% confidence interval of [0.902,0.918]. Compared with SVM, RF, and RF-CRF, OA increased by 15%, 6%, and 2.4%, respectively, and classification reliability increased by 17%, 6%, and 2%, respectively. The reason is that SVM and RF classify single pixels, which are inevitably misclassified even with the inclusion of textural information. CRF can use neighbourhood information to correct misclassified pixels, thereby, improving the classification accuracy. The comparison of the above results shows that the RF-Im_CRF model can further significantly reduce the noise generated in the random forest classification and improve the smoothness of images due to the correction capability of Im_CRF.

In order to analyse the classification accuracy relationship between each category, we give the experimental result data obtained in a single experiment, as shown in Table 3. In the absence of CRF, the 95% confidence interval of each class of random forest is basically between $[A + 2\%, A - 2\%]$. Where A represents the classification accuracy of each category. The Bootstrap Resampling method of the random forest causes each decision tree to use a different training subset, which leads to differences in classification performance across the trees. With a large number of decision trees, the random forest itself is more accurate than the SVM method, but it inevitably generates randomness, which results in slightly different classification results for each category. The number of test sets for each category is 150, which means that there are three different classification results for this category in the two experiments, and there will be a 2% difference.The classification effect is further improved by the CRF, resulting in a 95% confidence interval between $[A + 1\%, A - 1\%]$.

Table 3. Comparison of results of different classifiers.

Model		Water	High	Building	Low	Road
	Precision (%)	87	85	72	79	74
	Recall (%)	77	88	84	84	63
	F1-score (%)	82	86	78	81	70
	Precision (%)	98	92	79	85	78
RF	Recall (%)	95	93	91	81	72
	F1-score (%)	96	92	85	83	75
	Precision (%)	99	96	80	90	82
RF-CRF	Recall (%)	95	95	93	88	75
	F1-score (%)	97	95	86	89	78
	Precision (%)	100	97	84	93	88
RF-Im_CRF	Recall (%)	95	96	97	89	84
	F1-score (%)	97	96	90	91	86

It can be seen that the four models are more accurate in classifying water, high vegetation, and low vegetation than buildings and roads. The reason is that buildings have high complexity in both spectrum and structural characteristics, while roads are more challenging to identify due to low image resolution, a narrow area, and a susceptibity to factors, such as street trees. Among the two, roads are the most difficult to identify and the most error-prone category. This is because roads are mostly between buildings including the boundary between the road and the building that will blur the road with low image resolution. Moreover, the backscattering characteristics of buildings in SAR image can obscure the road to a certain extent, which has a negative impact on classification and makes roads more likely to be misclassified as buildings. At the same time, in the mixed area of multi-category features, low-resolution images significantly increase the complexity of categories, which makes the boundaries between categories difficult to distinguish. Therefore, how to effectively select feature quantities or improve image resolution to enhance the classification effect of buildings and roads, and make more precise distinctions to mixed regions will become the following research focus.

In terms of the model's operational efficiency, since the model proposed in this work needs to use neighbourhood information, this means that neighbourhood pixels must be classified as well. On the contrary, the original random forest classifier does not need to classify neighbourhood pixels. Therefore, the computational amount in the calculation process for this model is significantly higher than the one required for simpler classifiers, such as SVM or random forest. The evaluation of computing efficiency and the possible improvements of the algorithm from the computational point-of-view are in progress and will be the subject of the follow-up work.

5.2.3. Analysis of Feature Importance

This article also extracted the feature importance of each feature vector in the above ten experiments and took the average to get the results shown below.

As shown in Tables 4 and 5 and Figure 6, the feature importance of Freeman decomposition and spectral features are higher than others in the random forest classification. For the individual feature vectors, the volume scattering component in Freeman decomposition has the highest feature importance, which is followed by the blue component of spectral features. Nevertheless, the difference between the components of the spectral characteristics is not significant. This is because the volume scattering component is generally higher in the Freeman decomposition than the surface scattering and dihedral scattering for all targets except water. In water targets, these three components are small, and the scattering properties of road targets are similar to water under ideal conditions. Therefore, the volume scattering component has a good basis for judging the water area or road. Therefore, the body scattering has the highest feature importance. The recognition rate is not as ideal in water areas because of the complex and narrow environment in which roads are located.

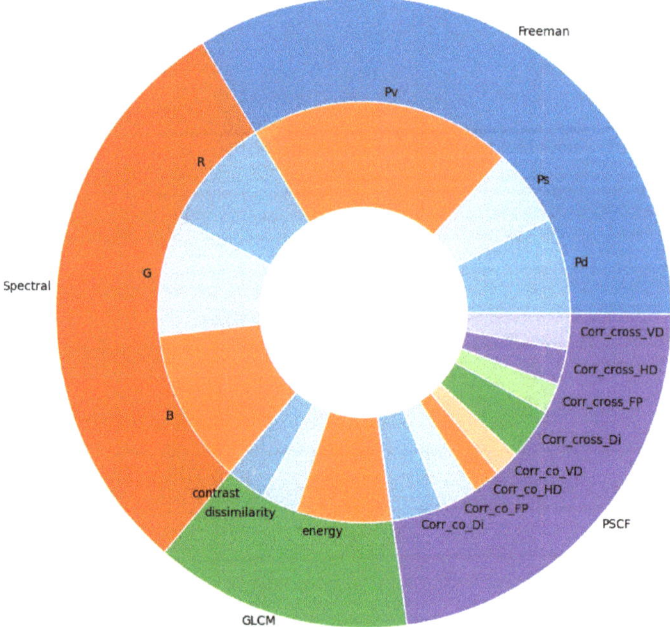

Figure 6. Feature importance ring graph.

Table 4. Feature importance of four characteristic components.

Feature	Freeman	Spectral	GLCM	PSCF
η (%)	33.78	30.03	13.44	22.72

Table 5. Feature importance of each feature.

Class	Pd	Ps	Pv	R	G	B	G_1	G_2	G_3	P_1	P_2	P_3	P_4	P_5	P_6	P_7	P_8
Im (%)	7.35	5.91	20.53	8.94	9.15	11.96	3.11	2.84	7.49	3.93	2.93	2.14	2.01	3.95	2.30	2.74	2.74

G_1 = contrast, G_2 = dissimilarity, G_3 = energy, P_1 = Corr_co_Di, P_2 = Corr_co_FP, P_3 = Corr_co_HD, P_4 = Corr_co_VD, P_5 = Corr_cross_Di, P_6 = Corr_cross_ FP, P_7 = Corr_cross_ HD, P_8 = Corr_cross_VD.

Except for energy, the GLCM and PSCF have similar proportions, while PSCF components are higher, so the η value is relatively high. The feature importance reflects the contribution degree of each feature in the classification. The randomness of random forest also impacts the feature importance. Therefore, the 95% confidence interval of four characteristic components is between [A−1%, A+1%]. Using such a contribution degree as the weight in the CRF pairwise potential function clarifies the spatial relationship between the target and the neighbourhood and improves classification accuracy.

6. Conclusions

Relying on the unique advantages of CRF in spatial context feature modelling and classification, this paper established a pixel-based RF-Im_CRF model for classification based on various feature information, such as spectrum, texture, and polarization. The experiments and analyses were carried out using polarized SAR and optical images of Nanjing area as data. The results show that the fusion of multi-source image data improves the classification accuracy. The RF-Im_CRF model with multiple features proposed in this paper further improves the classification accuracy to more than 94%, which increases by 6% when compared with the random forest classifier. Therefore, the RF-Im_CRF model has good performance in the fusion classification of polarized SAR and optical images and can be used as a fusion classification method for heterogeneous images.

Author Contributions: All the authors made a significant contribution to the work. Conceptualization, Y.K. and B.Y. Methodology, Y.K. and B.Y. Software, B.Y. Validation, Y.K., B.Y., and Y.L. Formal analysis, B.Y. Writing—original draft preparation, B.Y. Writing—review and editing, Y.K. and Y.L. Supervision, H.L. Project administration, X.P. All authors have read and agreed to the published version of the manuscript.

Funding: This research was funded by the National Natural Science Foundation of China (No. 61501228), the Natural Science Foundation of Jiangsu (No. BK20140825), the Aeronautical Science Foundation of China (No.20152052029, No.20182052012), Basic Research (No. NS2015040), and the National Science and Technology Major Project (2017-II-0001-0017).

Conflicts of Interest: The authors declare no conflict of interest.

References

1. Aswatha, S.M.; Mukherjee, J.; Biswas, P.K.; Aikat, S. Unsupervised classification of land cover using multi-modal data from mul-ti-spectral and hybrid-polarimetric SAR imageries. *Int. J. Remote Sens.* **2020**, *41*, 5277–5304. [CrossRef]
2. Hall, D.K.; Riggs, G.A.; Salomonson, V.V. Development of methods for mapping global snow cover using moderate resolution imaging spectroradiometer data. *Remote Sens. Environ.* **1995**, *54*, 127–140. [CrossRef]
3. Wu, J.; Zhang, Q.; Li, A.; Liang, C.Z. Historical landscape dynamics of Inner Mongolia: Patterns, drivers, and impacts. *Landsc. Ecol.* **2015**, *30*, 1579–1598. [CrossRef]
4. Useya, J.; Chen, S. Exploring the Potential of Mapping Cropping Patterns on Smallholder Scale Croplands Using Sentinel-1 SAR Data. *Chin. Geogr. Sci.* **2019**, *29*, 626–639. [CrossRef]
5. Neetu; Ray, S.S. Evaluation of different approaches to the fusion of Sentinel -1 SAR data and Resourcesat 2 LISS III optical data for use in crop classification. *Remote Sens. Lett.* **2020**, *11*, 1157–1166. [CrossRef]
6. Malthus, T.J.; Madeira, A.C. High resolution spectroradiometry: Spectral reflectance of field bean leaves infected by Botrytis fabae. *Remote Sens. Environ.* **1993**, *45*, 107–116. [CrossRef]
7. Sun, J.; Mao, S. River detection algorithm in SAR images based on edge extraction and ridge tracing techniques. *Int. J. Remote Sens.* **2011**, *32*, 3485–3494. [CrossRef]
8. Pohl, C.; Genderen, J.L. Review Article Multisensor Image Fusion in Remote Sensing: Concepts, Methods and Applications. *Int. J. Remote Sens.* **1998**, *19*, 823–854. [CrossRef]

9. Su, R.; Tang, Y. Feature Fusion and Classification of Optical-PolSAR Images. *Geomat. Spat. Inf. Technol.* **2019**, *42*, 51–55.
10. Zhang, L.; Zou, B.; Zhang, J.; Zhang, Y. Classification of Polarimetric SAR Image Based on Support Vector Machine Using Multiple-Component Scattering Model and Texture Features. *EURASIP J. Adv. Signal Process.* **2010**, *2010*, 960831. [CrossRef]
11. Ojala, T.; Pietiainen, M.; Harwood, D. A Comparative Study of Texture Measures with Classification Based on Feature Distributions. *Pattern Recognit.* **1996**, *29*, 51–59. [CrossRef]
12. Haralick, R.M.; Shanmugam, K.; Dinstein, I. Textural Features for Image Classification. In *IEEE Transactions on Systems, Man, and Cybernetics*; IEEE: New York, NY, USA, 1973; Volume SMC-3, pp. 610–621. [CrossRef]
13. Dong, J. *Statistical Analysis of Polarization SAR Image Features and Research on Classification Algorithm*; Wuhan University: Wuhan, China, 2018.
14. Freeman, A.; Durden, S.L. A three-component scattering model for polarimetric SAR data. *IEEE Trans. Geosci. Remote Sens.* **1998**, *36*, 963–973. [CrossRef]
15. Sato, A.; Yamaguchi, Y.; Singh, G.; Park, S.E. Four-Component Scattering Power Decomposition with Extended Volume Scattering Model. *IEEE Geosci. Remote Sens. Lett.* **2012**, *9*, 166–170. [CrossRef]
16. Attarchi, S. Extracting impervious surfaces from full polarimetric SAR images in different urban areas. *Int. J. Remote Sens.* **2020**, *41*, 4644–4663. [CrossRef]
17. Phartiyal, G.S.; Kumar, K.; Singh, D. An improved land cover classification using polarization signatures for PALSAR 2 data. *Adv. Space Res.* **2020**, *65*, 2622–2635. [CrossRef]
18. Breiman, L. *Manual on Setting Up, Using, and Understanding Random Forests V3.1 [EB/OL]*; Statistics Department University of California Berkley: Berkley, CA, USA, 2002.
19. Du, P.; Samat, A.; Waske, B.; Liu, S.; Li, Z. Random Forest and Rotation Forest for fully polarized SAR image classification using po-larimetric and spatial features. *ISPRS J. Photogramm. Remote Sens.* **2015**, *105*, 38–53. [CrossRef]
20. Sutton, C.; Mccallum, A. An Introduction to Conditional Random fields. *Found. Trends Mach. Learn.* **2010**, *4*, 267–373. [CrossRef]
21. Zhong, Y.; Jia, T.; Ji, Z.; Wang, X.; Jin, S. Spatial-Spectral-Emissivity Land-Cover Classification Fusing Visible and Thermal Infrared Hyperspectral Imagery. *Remote Sens.* **2017**, *9*, 910. [CrossRef]
22. Wang, L.; Zhou, X.; Zhu, X.; Dong, Z.; Guo, W. Estimation of biomass in wheat using random forest regression algorithm and remote sensing data. *Crop J.* **2016**, *4*, 212–219. [CrossRef]
23. Harold, M. The Kennaugh matrix. In *Remote Sensing with Polarimetric Radar*, 1st ed.; John Wiley & Sons: Hoboken, NJ, USA, 2007; pp. 295–298.
24. Lee, J.S.; Grunes, M.R.; Boerner, W.M. Polarimetric property preservation in SAR speckle filtering. In *Proceedings of SPIE 3120, Wideband Interferometric Sensing and Imaging Polarimetry*; Mott, H., Ed.; SPIE: San Diego, CA, USA, 1997; pp. 1–7.
25. Lee, J.S.; Pottier, E. Electromagnetic vector scattering operators. In *Polarimetric Radar Imaging: From Basics to Applications*, 1st ed.; Thompson, B.J., Ed.; CRC Press: New York, NY, USA, 2009; pp. 92–98.
26. Menze, B.; Kelm, B.; Masuch, R.; Himmelreich, U.; Bachert, P.; Petrich, W.; Hamprecht, F. A comparison of random forest and its Gini importance with standard chemometric methods for the feature selection and classification of spectral data. *BMC Bioinform.* **2009**, *10*, 213. [CrossRef] [PubMed]
27. Altmann, A.; Toloşi, L.; Sander, O.; Lengauer, T. Permutation importance: A corrected feature importance measure. *Bioinformatics (Oxf. Engl.)* **2010**, *26*, 1340–1347. [CrossRef] [PubMed]
28. Strobl, C.; Boulesteix, A.L.; Kneib, T.; Augustin, T.; Zeileis, A. Conditional variable importance for random forests. *BMC Bioinform.* **2008**, *9*, 307. [CrossRef] [PubMed]
29. Pedregosa, F.; Varoquaux, G.; Gramfort, A.; Michel, V.; Thirion, B.; Grisel, O.; Blondel, M.; Prettenhofer, P.; Weiss, R.; Dubourg, V.; et al. Scikit-learn: Machine Learning in Python. *J. Mach. Learn. Res.* **2011**, *12*, 2825–2830.

Article

Oil Spill Detection in SAR Images Using Online Extended Variational Learning of Dirichlet Process Mixtures of Gamma Distributions

Ahmed Almulihi [1], Fahd Alharithi [1], Sami Bourouis [1,*], Roobaea Alroobaea [1], Yogesh Pawar [2] and Nizar Bouguila [2]

1 College of Computers and Information Technology, Taif University, P.O. Box 11099, Taif 21944, Saudi Arabia; a.almulihi@tu.edu.sa (A.A.); f.alshalawi@tu.edu.sa (F.A.); r.robai@tu.edu.sa (R.A.)
2 The Concordia Institute for Information Systems Engineering (CIISE), Concordia University, Montreal, QC H3G 1T7, Canada; yogesh.pawar@mail.concordia.ca (Y.P.); nizar.bouguila@concordia.ca (N.B.)
* Correspondence: s.bourouis@tu.edu.sa

Abstract: In this paper, we propose a Dirichlet process (DP) mixture model of Gamma distributions, which is an extension of the finite Gamma mixture model to the infinite case. In particular, we propose a novel online nonparametric Bayesian analysis method based on the infinite Gamma mixture model where the determination of the number of clusters is bypassed via an infinite number of mixture components. The proposed model is learned via an online extended variational Bayesian inference approach in a flexible way where the priors of model's parameters are selected appropriately and the posteriors are approximated effectively in a closed form. The online setting has the advantage to allow data instances to be treated in a sequential manner, which is more attractive than batch learning especially when dealing with massive and streaming data. We demonstrated the performance and merits of the proposed statistical framework with a challenging real-world application namely oil spill detection in synthetic aperture radar (SAR) images.

Keywords: Dirichlet process; infinite mixture models; Gamma distribution; variational inference; online setting; oil spill detection; synthetic aperture radar images

Citation: Almulihi, A.; Alharithi, F.; Bourouis, S.; Alroobaea, R.; Pawar, Y.; Bouguila, N. Oil Spill Detection in SAR Images Using Online Extended Variational Learning of Dirichlet Process Mixtures of Gamma Distributions. *Remote Sens.* **2021**, *13*, 2991. https://doi.org/10.3390/rs13152991

Academic Editors: Monidipa Das, Soumya K. Ghosh, Vemuri M. Chowdary, Pabitra Mitra and Santosh Rijal

Received: 8 June 2021
Accepted: 26 July 2021
Published: 29 July 2021

1. Introduction

The use of statistical machine learning has proliferated in many fields, especially to solve a broad range of problems ranging from signal processing, speech recognition, to geosciences and remote sensing where strong models are needed to apply statistical methodology. In the case of geosciences and remote sensing, for instance, statistical machine learning techniques have been deployed successfully in a variety of problems and applications in many parts of the earth system and beyond [1]. In particular, images modeling (e.g., SAR images) has received much attention due to its importance and applications in real world nature tasks related to land, climate, disturbance attribution, vegetation dynamics, urbanization, etc.

Among the probabilistic generative models, the so-named finite mixtures have been successfully applied due to their capability to represent large-scale complex probability densities and to offer a principled way for analyzing missing data [2,3]. Mixture models provide, in general, a formal approach to unsupervised learning and allow, in particular, to select the optimal number of clusters for a given dataset. This fact has been largely detailed in the literature (see, for example, [4,5]). This growing interest has led to developing several fascinating and flexible mixture models such as Gaussian-based mixture models (GMM) which have became popular even though they are not the most appropriate for fitting complex non-Gaussian shapes [6,7]. To deal with conventional GMM limitations, many other alternatives, such as Gamma (GaMM) mixtures [8–11], have shown to perform significantly better than GMM [12] thanks to its compact analytical form which is able to cover long-tailed distributions and to approximate data with outliers. Thus, motivated by the flexibility and good performance obtained with Gamma distribution, we will focus here on investigating Gamma-based mixture model for SAR images classification. We are mainly motivated by the excellent results that Gamma mixture has provided, thanks to its flexibility, for SAR images analysis in many applications such as target detection and discrimination, target recognition and surface classification, oil spill detection, noise reduction, etc. [10]. In this paper, we will focus mainly on oil spill detection

The most challenging problem within finite mixture models is the estimation of the number of clusters that best describes the data without over- or under-fitting [13,14]. In the statistical learning context, this problem is solved using frequentist approach (i.e., maximum likelihood (ML)) within some criteria (ex. Akaike's Information Criterion, Minimum Description Length, Minimum Message Length, etc) [15,16]. It is noteworthy that the evaluation of these criteria for many clusters using ML method is very costly in terms of calculation. In addition, all parameters are supposed fixed and the inference process is based mainly on the likelihood of data which leads to convergence isssues. An alternative way to tackle the issue of selecting accurately the number of clusters is via nonparametric Bayesian inference using for instance Dirichlet process (DP) [17]. In this case, the number of clusters may increase as more data are observed. This property makes DP extremely useful in exploratory data analysis. Thus, the assumption of an infinite number of components allows to avoid the problems of over- and under-fitting. Dirichlet processes (DP) mixtures have become a popular choice for various machine learning applications thanks to effective sampling techniques such as Markov chain Monte Carlo (MCMC) [18,19]. Despite the fact that MCMC yields good performance, it is frequently limited to small-scale problems and computationally intensive [20].

An interesting alternative, to both frequentist and Bayesian methods, which has provided promising performance, is variational Bayes learning [15,21]. Variational inference has the advantage to find optimal approximate posterior distributions by minimizing Kullback–Leibler (KL) divergence, or as maximizing evidence lower bound. Recently, an extended variational inference (EVI) was proposed [8] and has shown to be efficient for minimizing the KL divergence and for tackling the estimation problem. In this work, we go a step further by developing an infinite mixture model based on Gamma distribution via Dirichlet process prior, and then we propose to exploit the merits found recently by the extended variational framework [8] to learn the developed mixture model (InGaMM-eV) in an online manner. Furthermore, it is possible to estimate all parameters in closed forms. Moreover, compared to batch algorithms, online learning is more effective and helpful especially when processing big and streaming data [22] which can be crucial in SAR images analysis to allow continuous monitoring of the earth's surface. It is noteworthy also that many SAR satellite missions have accumulated repeated observations over the last decades and processing these data in an online manner could offer ease of use and solutions to some challenging problems (e.g., change detection [23]). Thus, an effective online extended variational framework of Dirichlet process mixtures of Gamma distributions is developed using stick-breaking representation. As a result, the number of clusters is selected appropriately, the model's parameters are learned in a closed form, and the issue of under-fitting is solved by deriving a model with an unlimited complexity.

The rest of this manuscript is presented as follows. We review some relevant works related to oil spill detection in Section 2. The details of extending the finite Gamma mixture to infinite case are given in Section 3. The principles of our implemented nonparametric variational learning algorithm of infinite Gamma mixture are provided in Sections 4 and 5. Section 6 is devoted to discuss the results obtained from experiments. Finally, the paper is concluded with some future works.

2. Related Research Work

Oil pollution is a major ocean disaster and environmental threat to coastal ecosystems which has been recently highlighted by several tankers accidents around the world.

Accidents on offshore oil platforms, refineries, and pipeline can cause serious oil spills. However, these accidents represent only 5% of the total oil pollution worldwide, and 95% are caused by illegal discharges by ships that prefer to dispose, cheaply, of oil residues in their tanks (according to many studies such as the European Space Agency) [24–26]. Oil pollution may result from several sources such as industrial discharges, oil production, natural oil seepage, and urban runoff. Natural slicks are of bacterial or biological decomposition or geological origin. Oil spills can devastate naval life as well as harm humans and animals by reducing dramatically air-sea exchanges processes, such as surface evaporation. Oil spills are then of great public, political and scientific concern. Therefore, there is an urgent need to monitor and detect oil spills on ocean so as to facilitate government decision making. The detection of these oil spills is considered an important and challenging problem to effectively conduct countermeasures. An effective approach is the use of satellites which provide radar images of the sea surface (500 × 500 km^2 in a single image). Satellites radar images supply an occasion to monitor coastal waters day and night, regardless of weather conditions allowing an early warning of oil spills. Moreover, satellite detection is well adapted to this kind of problems by producing images of difficult access areas [24]. Among different satellite imagery technologies, active microwave sensors such as synthetic aperture radar (SAR), has been frequently investigated for remote sensing of oil pollution [27]. The synthetic aperture radar emits and receives radio wave in order to acquire a representation of the target scene. Detecting oil spill in SAR images (as shown in Figure 1) is very complex procedure that involves many steps [26].

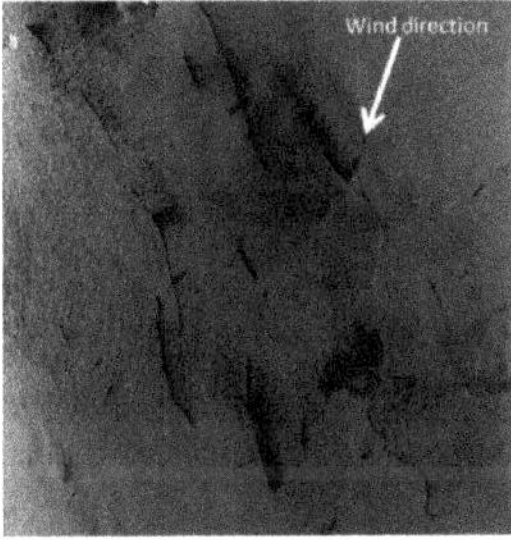

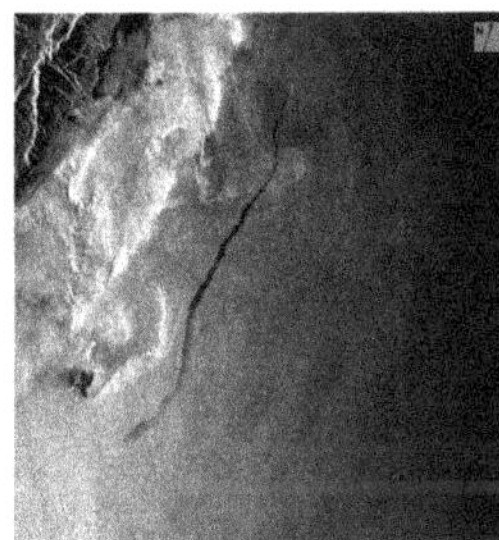

Figure 1. SAR image obtained by the European Remote Sensing satellite ERS-2 on April 1997 over the South China Sea (left image) and SAR image obtained by the ERS-1 satellite on May 1994 over Pacific Ocean east of Taiwan (right image). These images (area: 100 km × 100 km) showing an oil spill [28].

For several decades, extensive works have been provided [27,29,30] to distinguish oil slicks from natural biogenic slicks via analyzing satellite radar images. Most of conventional oil slick (or dark objects) detection procedures are carried out in three steps: (1) a pre-segmentation of dark spot, (2) the extraction of dark spot feature, and (3) a classification step of these dark spots. Some early and recent review articles summarize different oil slick detection methods [26,28,31]. These reviews state that most methods are based on using statistical patterns to discriminate between oil slicks and look-alikes under varying conditions. They conclude also that the automatic and accurate discrimination between oil spills and look-alikes is a challenging problem and need more investigations in the future. On the other side, a lot of efforts have been devoted to apply classic classifiers and descriptive statistical approaches learned from training data [25,30,32–34]. These works rely on highly trained human operators to asses and verify each region in a given SAR image. In [33], authors proposed a one-class based approach for image classification to detect oil-spill. First of all, a preprocessing step is used to identify related areas to oil spills. A feature selection step to select relevant features is also performed given that the contrast between spill's region and the surrounding regions depends on the type and amount of oil and other environmental factors (i.e., wave height, wind speed, and sea). Finally, a one-class classifier is used to detect oil spills. A geometric level-set based segmentation method of oil spills and illegal oil discharges was developed in [35]. According to this work the regions in SAR images can be classified into pure oil spills or look-alikes on the basis of the following measurements: orientation, area, shape complexity, perimeter, eccentricity, and mean border gradient. In [36], a region-based method was also proposed. It involves both conventional detection theory and image segmentation techniques (such as N-nearest-neighbor) to have more accurate results. In [37], authors developed an adaptive thresholding-based algorithm to classify each slick as oil or look-alike. Here, involved features are derived from shape (slick complexity, width, area, moment), slick surroundings, contrast (slick local contrast, border gradient, smoothness contrast), and slick homogeneity. Their algorithms have been trained on two datasets, namely Radarsat and Envisat Advanced Synthetic Aperture Radar (ASAR) images. Fuzzy classifiers have been also used in [38] to identify all possible oil spills (dark patterns) in SAR images. A set of operations based on the fuzzy theory are used to establish the likeness of each candidate to be an oil spill or not. In the last few years, artificial neural network algorithms have been broadly applied in the context of remote sensing image segmentation and classification. Indeed, authors in [39–43] proposed different neural network-based methods (like CNN and Deep NN) in order to improve oil spill detection and classification. Some other notable interesting CNN-based oil spill detection and classification frameworks include the works in [44,45].

While considerable progress has been made in this field over the past few years, designing more robust tools still needs wide amounts of specialized knowledge and manual work. The goal here is to propose a method based on a nonparametric Bayesian model (infinite model) as well as to learn it using variational inference. Our main contributions are summarized as follow: First, we start by extending the finite Gamma mixture to the infinite case via a nonparametric Dirichlet process prior such that the problem of selecting the suitable number of clusters is solved fashionably. Then, we investigate the developed approach for remote sensing image classification. Indeed, after extracting effective features as in [46], we shall focus on modelling and classifying oil spills and other similar sea surface features using the infinite mixture model. The merits of our approach have been demonstrated using real datasets.

3. Statistical Model Specification

In this section, we present our developed variational learning approach based on the infinite Gamma mixture model.

3.1. Finite Gamma Mixture Model

Let's denote by $\mathcal{Y}$ our observed data such as $\mathcal{Y} = \{\vec{Y}_1, \dots, \vec{Y}_N\}$, where each $\vec{Y}_i = (Y_{i1}, Y_{i2}, \dots, Y_{iD})$ is a D -dimensional positive vector. These feature vectors are supposed to be drawn from a mixture of Gamma distributions with parameter Θ. Let M denotes the number of mixture's components. $\vec{Y}_i$ $(i = 1, \dots, N)$ are independent and identically distributed (iid). The density function of multi-dimensional Gamma distribution is defined as follows:

$$p(\vec{Y}_i \mid \theta) = \prod_{d=1}^{D} \frac{(\beta_d)^{\alpha_d} Y_{id}^{\alpha_d - 1} e^{-\beta_d Y_{id}}}{\Gamma(\alpha_d)} \tag{1}$$

where $\theta = \{\alpha_d, \beta_d\}$ is the set of parameters of the distribution such that α_d denotes the shape and β_d the location parameter. Here, $\Gamma(.)$ is the Gamma function which is given as: $\Gamma(x) = \int_0^\infty s^{x-1} e^{-s} ds$.

Suppose that the D-dimensional random vector $\vec{Y}_i$ (observed data) is drawn from a finite mixture of Gamma (GaMM) distributions and consisting of M components which is established to model the data with different shapes. The probability density function (pdf) of a GaMM is then given as:

$$p(\vec{Y} \mid \Theta) = p(\vec{Y} \mid \vec{\alpha}, \vec{\beta}, \vec{\pi}) = \prod_{i=1}^{N} \sum_{j=1}^{M} \pi_j p(\vec{Y}_i \mid \theta_j) \tag{2}$$

where $\Theta = \{\theta_1, \theta_2, \ldots, \theta_M, \pi_1, \ldots, \pi_M\}$. The parameters of the jth mixture component is represented by $\theta_j = \{\alpha_j, \beta_j\}$. π_j is the vector of the mixing weights subject to $0 \leqslant \pi_j \leqslant 1$, and $\sum_{j=1}^{M} \pi_j = 1$.

3.2. *Infinite Gamma Mixture Model*

The Dirichlet process (DP) is a stochastic process with a positive scaling factor and base distribution used in Bayesian nonparametric models of data, notably in infinite mixture models. The DP is an effective concept for various applications (for more details please refer to [47]). In this section we address the issue of assuming an infinite number of components. In order to solve properly this problem which is important for well describing the observed data without over- or under-fitting, we propose a Dirichlet process mixture of Gamma distributions. In other words, we construct our infinite model by following the principle of Dirichlet process (DP) through stick-breaking representation [48,49]. Thus, the number of components is intended to be infinite. In this case, let's denote G a Dirichlet process distributed with a base distribution H and a concentration parameter ψ. The construction of $G \sim DP(\psi, H)$ is defined as

$$\begin{aligned}
&\lambda \sim Beta(1, \psi) \\
&\Omega_j \sim H \\
&\pi_j = \lambda_j \prod_{s=1}^{j-1} (1 - \lambda_s) \\
&G = \sum_{j=1}^{\infty} \pi_j \delta_{\Omega_j}
\end{aligned} \tag{3}$$

where δ_{Ω_j} represents the Dirac delta measure centred at Ω_j. The proportions π_j are determined by cutting a unit length stick, regularly, into an infinite number of pieces such that $\sum_{j=1}^{\infty} \pi_j = 1$ and ψ is a real number. Consequently, the infinite mixture model of Gamma distributions $\mathcal{Y}$ is expressed as

$$p(\mathcal{Y} \mid \Theta) = p(\mathcal{Y} \mid \vec{\alpha}, \vec{\beta}, \vec{\pi}) = \prod_{i=1}^{N} \sum_{j=1}^{\infty} \pi_j p(\vec{Y}_i \mid \theta_j) \tag{4}$$

Subsequently, a latent variable $Z_i = (Z_{i1}, Z_{i2}, \ldots)$ is introduced for observed data $\mathcal{Y}$. These latent membership vectors are used to point out if the vector $\vec{Y}_i$ belongs to component j (Z_{ij} = 1) or not (Z_{ij} = 0). Now, the complete-data likelihood is expressed as

$$p(\mathcal{Y}, Z \mid \vec{\alpha}, \vec{\beta}, \vec{\pi}) = \prod_{i=1}^{N} \prod_{j=1}^{\infty} \pi_j^{z_{ij}} \left(p(\vec{Y}_i \mid \alpha_j, \beta_j) \right)^{z_{ij}} \tag{5}$$

According to the stick-breaking construction of DP (see Equation (3)), π_j can be expressed as a function of λ_j and after replacement, we have the following:

$$p(\mathcal{Z} \mid \vec{\lambda}) = \prod_{i=1}^{N} \prod_{j=1}^{\infty} \left[\lambda_j \prod_{s=1}^{j-1} (1 - \lambda_s) \right]^{z_{ij}} \tag{6}$$

The resulting complete-likelihood of the infinite Gamma mixture is finally expressed as (including latent variables):

$$p(\mathcal{Y}, Z \mid \vec{\alpha}, \vec{\beta}, \vec{\pi}) = \prod_{i=1}^{N} \prod_{j=1}^{\infty} \left[\lambda_j \prod_{s=1}^{j-1} (1 - \lambda_s) \right]^{z_{ij}} \left(p(\vec{Y}_i \mid \alpha_j, \beta_j) \right)^{z_{ij}} \tag{7}$$

4. Batch Variational Bayesian Learning

It is noteworthy that, when dealing with intractable models, variational inference is presented as a powerful deterministic alternative to approximate posteriors and likelihoods. In this section, we propose to develop a variational learning method to approximate inference for the DP, where the truncated stick-breaking construction [50] is applied to derive an approximate posterior and to estimate the model parameters. On the other side, we proceed by determining an approximation $Q(\Theta)$ for true posterior $p(\Theta \mid Y)$ such that $\Theta = \{Z, \alpha, \beta\}$. After that, we use the well-known KL divergence in order to reduce the difference between $Q(\Theta)$ and $p(\Theta \mid Y)$:

$$KL(Q \,||\, P) = \int Q(\Theta)\, ln\left(\frac{p(\Theta \mid \mathcal{Y})}{Q(\Theta)} \right) d\Theta \tag{8}$$

$$KL(Q \,||\, P) = ln(p(\mathcal{Y}) - \mathcal{L}(Q) \tag{9}$$

$$\mathcal{L}(Q) = \int Q(\Theta)\, ln\left(\frac{p(\mathcal{Y}, \Theta)}{Q(\Theta)} \right) d\Theta \tag{10}$$

KL divergence attains value of zero if we have $Q(\Theta) = p(\Theta \mid \mathcal{Y})$ (since As $KL(Q \,||\, P) \geq 0$). From Equation (9), it is possible to deduce that $\mathcal{L}(Q) \leq lnp(\mathcal{Y})$ and so $\mathcal{L}(Q)$ is a lower bound to $lnp(\mathcal{Y})$. However, it is difficult to solve the true posterior which cannot be directly estimated because of the complexity of calculation. We get around this matter by taking into account a restricted family of $Q(\Theta)$ that can be calculated [21]. In particular, the mean field theory [51] is adopted to factorize $Q(\Theta)$ into different tractable distributions such that $Q(\Theta) = \prod_{i=1} Q_i(\Theta_i)$. To maximize $\mathcal{L}(Q)$, we apply variational methodology with respect to each $Q_i(\Theta_i)$. Then, the optimal form of $Q_i(\Theta_i)$ denoted by $Q_s(\Theta_s)$ is given as

$$lnQ_s(\Theta_s) = \langle ln(p(\mathcal{Y}, \Theta) \rangle_{j \neq s} + const \tag{11}$$

where $\langle . \rangle_{j \neq s}$ is the expectation value of Q, with respect to all $Q_i(\Theta_i)$ excluding that case of $j = s$. It is noted that we have to take into account the truncation of the stick-breaking representation [49] to take advantage of the bound. Therefore, we take $\lambda_M = 1$ and $\pi_j = 0$ when $j > M$ which leads to $\sum_{j=1}^{M} \pi_j = 1$.

4.1. Prior Distributions for Parameters

To complete the probabilistic formulation, we have to place proper conjugate priors over the parameters λ, α and β. In particular, the Beta distribution is selected for the parameter λ (referring to Equation (3)) as follow

$$p(\lambda \mid \psi) = \prod_{j=1}^{\infty} Beta(1, \psi_j) = \prod_{j=1}^{\infty} \psi_j (1 - \lambda_j)^{\psi_j - 1} \tag{12}$$

Here, the hyperparameters of the Beta distribution is denoted by $\psi = (\psi_1, \psi_1, \ldots)$ [52]. Moreover, it is possible to assign a conjugate Gamma prior to ψ:

$$p(\psi) = \mathcal{G}(\psi \mid a, b) = \prod_{j=1}^{\infty} \frac{b_j^{a_j}}{\Gamma(a_j)} \psi^{a_j - 1} e^{-b_j \psi_j} \tag{13}$$

For α and β, a prior Gamma distribution is imposed for them as suggested in [8] which is reasonable given that α and β are positives and also Gamma density is assumed to be too flexible and simple distribution to be selected as prior.

$$p(\vec{\alpha}) = \mathcal{G}(\vec{\alpha} \mid \vec{u}, \vec{v}) = \prod_{j=1}^{\infty} \prod_{d=1}^{D} \mathcal{G}(\alpha_{jd} \mid u_{jd}, v_{jd}) \tag{14}$$

$$p(\vec{\beta}) = \mathcal{G}(\vec{\beta} \mid \vec{s}, \vec{t}) = \prod_{j=1}^{\infty} \prod_{d=1}^{D} \mathcal{G}(\beta_{jd} \mid s_{jd}, t_{jd}) \tag{15}$$

Following the graphical model in Figure 1, the resulting joint distribution is expressed as

$$\begin{aligned} p(\mathcal{Y}, \Theta) &= p(\mathcal{Y}, \mathcal{Z} \mid \vec{\alpha}, \vec{\beta}) p(\mathcal{Z} \mid \vec{\lambda}) p(\vec{\lambda} \mid \vec{\psi}) p(\vec{\psi}) p(\vec{\alpha}) p(\vec{\beta}) \\ &= \prod_{i=1}^{N} \prod_{j=1}^{\infty} \left[\lambda_j \prod_{s=1}^{j-1} (1 - \lambda_s) \right]^{z_{ij}} \left(p(\vec{Y}_i \mid \alpha_j, \beta_j) \right)^{z_{ij}} \\ &\times \prod_{j=1}^{\infty} \psi_j (1 - \lambda_j)^{\psi_j - 1} \\ &\times \prod_{j=1}^{\infty} \frac{b_j^{a_j}}{\Gamma(a_j)} \psi^{a_j - 1} e^{-b_j \psi_j} \\ &\times \prod_{j=1}^{\infty} \prod_{d=1}^{D} \mathcal{G}(\alpha_{jd} \mid u_{jd}, v_{jd}) \\ &\times \prod_{j=1}^{\infty} \prod_{d=1}^{D} \mathcal{G}(\beta_{jd} \mid s_{jd}, t_{jd}) \end{aligned} \tag{16}$$

4.2. Learning Algorithm

As explained at the beginning, the objective of this work is to approximate the true posterior $p(\Theta \mid Y)$ with a new tractable approximation denoted by $Q(\Theta)$. Furthermore, the optimal solution of variational learning is reached while maximizing the lower bound w.r.t $\Theta = \{Z, \lambda, \alpha, \beta\}$. The factorization of $Q(\Theta)$ (while taking into account the truncation M) leads to following parametric form which optimal solution is presented in Appendix A:

$$Q(\Theta) = \left[\prod_{i=1}^{N} \prod_{j=1}^{M} Q(Z_{ij}) \right] \left[\prod_{j=1}^{M} Q(\lambda_j) Q(\psi_j) \right] \left[\prod_{j=1}^{M} \prod_{d=1}^{D} Q(\alpha_{jd}) Q(\beta_{jd}) \right] \tag{17}$$

Once the optimal variational factors are in hand, the calculation of the lower bound $\mathcal{L}(Q)$ is then straightforward. Figure 2 presents a graphical model of the proposed infinite Gamma mixture model (inGaMM). Random variables are denoted by circles and hyperparameters are represented by rounded boxes. Then, the different steps of the implemented method are summarized in Algorithm 1.

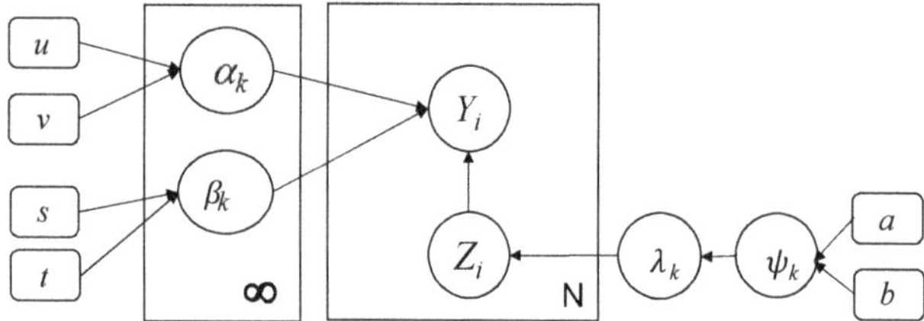

Figure 2. Graphical model of the developed variational infinite inGaMM. Random variables are denoted by circles and hyperparameters are represented by rounded boxes . Y is observed variable, Z is latent variable, large boxes are used for repeated process and the arrows show the conditional dependence between variables.

Algorithm 1: Batch variational learning approach for the inGaMM

```
Choose initial truncation for M.
Set initial values for hyperparameters u, v, s, t, a, b, c, d.
Initialize r_ij via k-means algorithm.
repeat
    Variational E-step :
    Estimate the expected values according to Equations (A5), (A9), (A12),
      and (A15).
    Variational M-step :
    Update the variational solution for the factor Q(Z) using Equation (A1).
    Update the variational solution for the factor Q(ψ) using Equation (A6).
    Update the variational solution for the factor Q(λ) using Equation (A7).
    Update the variational solution for the factor Q(α) using Equation (A10).
    Update the variational solution for the factor Q(β) using Equation (A13).
until Until convergence is reached
Calculate the expected value of λ_j using Equation (A9). Then estimate the mixing
  coefficients according to Equation (3).
Return the optimal number of components M_opt by eliminating the components
  with small mixing coefficients close to 0.
```

5. Online Variational Bayesian Learning

Early warning and immediate detection of oil spills has many advantages such as immediate response and reducing damage to the environment. The development of real-time monitoring and detection system is of great importance in order to minimize the volume of oil spilled. To address this problem, we propose to develop an online learning approach which is being commonly used in many other areas especially when data points are continuously arriving over time [53]. The online setting is particularly useful for incrementally training the system by feeding instances of data sequentially. It also has the benefit of making the learning process easier and faster than batch mode.

In what follows, we extend the batch variational method (presented in previous section) for unsupervised SAR images classification to an online setting. This process requires updating the model's parameters incrementally without degrading its efficiency and flexibility. To determine the lower bound, we suppose that we have at time t a fixed set of observed data. At time $t+1$, a new SAR image Y_{N+1} comes out and is added to the dataset, hence, the mixtures' parameters have to be updated accordingly. Thus, in online setting, the lower bound at time t is expressed as in [54]:

$$\mathcal{L}^t(Q) = \frac{N}{t}\sum_{i=1}^{t}\int Q(\Omega)d\Omega \sum_{Z_i} ln\left[\frac{p(\vec{Y}_i, \vec{Z}_i \mid \Omega)}{Q(\vec{Z}_i)}\right] + \int Q(\Omega) ln\left[\frac{p(\Omega)}{Q(\Omega)}\right] d\Omega \tag{18}$$

where $\Omega = \{\alpha, \beta\}$.

Let's suppose that we already observed $\{\vec{Y}_1, \dots, \vec{Y}_{(t-1)}\}$ and then a new data point $\vec{Y}_t$ is coming. Therefore, $\mathcal{L}^t(Q)$ is maximized w.r.t $Q(\vec{Z}_t)$, such that $Q(\alpha)$, $Q(\lambda)$ and $Q(\beta)$ are set to $Q^{t-1}(\alpha)$, $Q^{t-1}(\lambda)$ and $Q^{t-1}(\beta)$, respectively. We adopt a truncation technique with value M which gives [49]:

$$Q(\vec{Z}_t) = \prod_{j=1}^{M} r_{tj}^{Z_{tj}} \tag{19}$$

$$r_{tj} = \frac{\rho_{tj}}{\sum_{j=1}^{M} \rho_{tj}} \tag{20}$$

Then, $\mathcal{L}^t(Q)$ is maximized w.r.t $Q(\alpha)$, $Q(\lambda)$ and $Q(\beta)$ while keeping $Q(\vec{Z}_t)$ fixed.

$$Q^{(t)}(\vec{\alpha}) = \prod_{j=1}^{M} \prod_{d=1}^{D} \mathcal{G}(\alpha_{jd}^{(t)} \mid u_{jd}^{*(t)}, v_{jd}^{*(t)}) \tag{21}$$

$$Q^{(t)}(\vec{\beta}) = \prod_{j=1}^{M} \prod_{d=1}^{D} \mathcal{G}(\beta_{jd}^{(t)} \mid s_{jd}^{*(t)}, t_{jd}^{*(t)}) \tag{22}$$

$$Q^{(t)}(\lambda) = \prod_{j=1}^{M} Beta(\lambda_j^{(t)} \mid c_j^{(t)}, d_j^{(t)}) \tag{23}$$

where

$$\begin{aligned}
u_{jd}^{*(t)} &= u_{jd}^{*(t-1)} + \rho_t \Delta u_{jd}^{*(t)} \\
v_{jd}^{*(t)} &= v_{jd}^{*(t-1)} + \rho_t \Delta v_{jd}^{*(t)} \\
s_{jd}^{*(t)} &= s_{jd}^{*(t-1)} + \rho_t \Delta s_{jd}^{*(t)} \\
t_{jd}^{*(t)} &= t_{jd}^{*(t-1)} + \rho_t \Delta t_{jd}^{*(t)} \\
c_{jd}^{*(t)} &= c_{jd}^{*(t-1)} + \rho_t \Delta c_{jd}^{*(t)} \\
d_{jd}^{*(t)} &= d_{jd}^{*(t-1)} + \rho_t \Delta d_{jd}^{*(t)}
\end{aligned} \tag{24}$$

Δ is the natural gradient of each hyperparameter in the previous equation. ρ_t denotes the learning rate [55] expressed by following equation:

$$\rho_t = (\eta_0 + t)^{-\epsilon} \tag{25}$$

where $\epsilon \in [0.5, 1]$ and $\eta \geq 0$. This helps to guarantee convergence [55]. Please note that the expectation in the above mentioned equations are obtained with same manner as for the case of batch setting in the previous section and as in [56]. Since the online learning framework can be considered as a stochastic approximation algorithm, the convergence is ensured as prove in [53]. The proposed and developed online variational algorithm is presented in Algorithm 2.

Algorithm 2: Proposed online algorithm for inGaMM

Select initial truncation level M.
Set initial values for hyperparameters
Initialize r_{tj} via k-means algorithm.
repeat
    Variational E-step :
    Update the variational solution for Equation (19).
    Variational M-step :
    Compute the learning rate using Equation (25).
    Calculate the hyperparameters using Equation (24).
    Update the variational solutions for all factors $Q^{(t)}(\alpha)$, $Q^{(t)}(\beta)$ and $Q^{(t)}(\lambda)$ using Equations (21)–(23).
    Repeat the variational E-step and V-step until new data is observed
until *for t = 1 to N*

6. Experimental Results

6.1. Data Sets

The main objective of this section is to investigate our developed online extended variational learning framework of Dirichlet process mixture of Gamma distributions to detect oil spills in several SAR images. The second objective is to compare the performance of the proposed statistical framework with other methods from the state-of-art. First, it should be noted that one of the challenges is the lack of already common data sets for oil spill detection and this problem has been addressed by many relevant research communities such as [57,58]. Very limited data sets have been proposed in the literature, and therefore, it is too difficult to compare between published results since each method uses different data sets with different settings. In this work, we are essentially concerned with two challenging SAR databases. The first data set is the SAR images containing oil spills collected via the European Space Agency (ESA) database [40] which is composed of 1112 images with 5 different classes: Land, Look-alike, oil-spill, ships, and sea surface. The second one is a labelled SAR dataset taken from Sentinel-1 wave mode (TenGeoP-SARwv) [59] which includes 40,553 images with 10 different geophysical phenomena such as Pure Ocean Waves (F), Wind Streaks (G), Micro Convective Cells (H), Rain Cells (I), Biological Slicks (J), Sea Ice (K), Iceberg (L), Low Wind Area (M), Atmospheric Front (N), and Oceanic Front (O). Figures 3 and 4 show examples of images from these two datasets, respectively. For experiments, we randomly select half of the dataset as the training set and the rest for testing. In order to quantify how well SAR images are classified, we report the results in terms of average accuracy metric and false positive rate (FPR).

Modeling and classifying SAR requires powerful statistical models to represent their content (ex. color, texture). In this work we shall focus on the problem of SAR images modeling and classification via extracting local features that describe accurately input images. Indeed, feature extraction step is a part of the dimensionality reduction process that has been broadly studied in the past. It has an important role in many computer vision applications since it helps identifying the most discriminating characteristics, reducing ambiguity and enhancing the performance. However, the presence of speckle noise in synthetic aperture radar (SAR) images, as well as low-resolution between regions (surfaces) and poor contrast, make extracting relevant features too difficult. Thus, if the representative features are well extracted, then we can correctly interpret and classify images. Extracting local features from grey-scale images is a well-studied step in the fields of image processing and computer vision and various comparative measures have been studied for many years. The study of prior techniques is not within the scope of this paper. However, we suggest applying two successful methods of features extraction. The first one is based on imageNet pretrained deep learning model (resnet50) [60]. The flowchart diagram for extracting features using resnet50 is given in Figure 5. For each SAR image in the flowchart, we first apply different image processing operations like adjusting contrast value, thresholding, object edge detection by blurring noise and small objects. After this step, based on the number of detected dark spots, we extract different features including geometrical characteristics and texture of the object. Finally, we store the extracted features for the model evaluation. In the second approach, we extract a number of features based on geometrical characteristics, physical behavior, and those related to oil spill context of the dark formations as described in [61]. After extracting features, we applied principal component analysis (PCA) to reduce dimensionality of extracted datasets features.

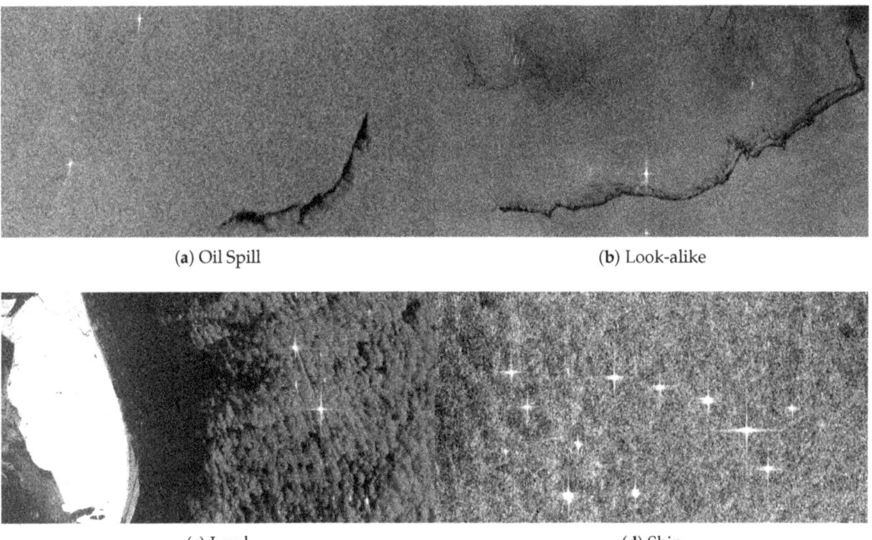

Figure 3. Dataset-1: Samples of SAR images from the European Space Agency (ESA) dataset [40]. (**a**) OilSpill, (**b**) Look-alike, (**c**) Land, (**d**) Ship.

6.2. Results and Discussion

Next, we apply our online extended variational algorithm (Section 5) over the extracted features. Thus, each image is represented by an infinite Gamma mixture model. We average the results over 30 runs to evaluate and compute the final performance. Tables 1 and 2 show the average classification accuracy and false positive rate (FPR) of our InGaMM-eV model. They are obtained with different classes in both datasets and by using two features extraction methods. Indeed, we considered a first experiment where the goal was to distinguish between oil spills versus the rest and a second one where the goal is to categorize some classes from each data set (4 categories are taken from the first data set and 9 from the second one). The testing data is assumed to arrive sequentially in an online mode.

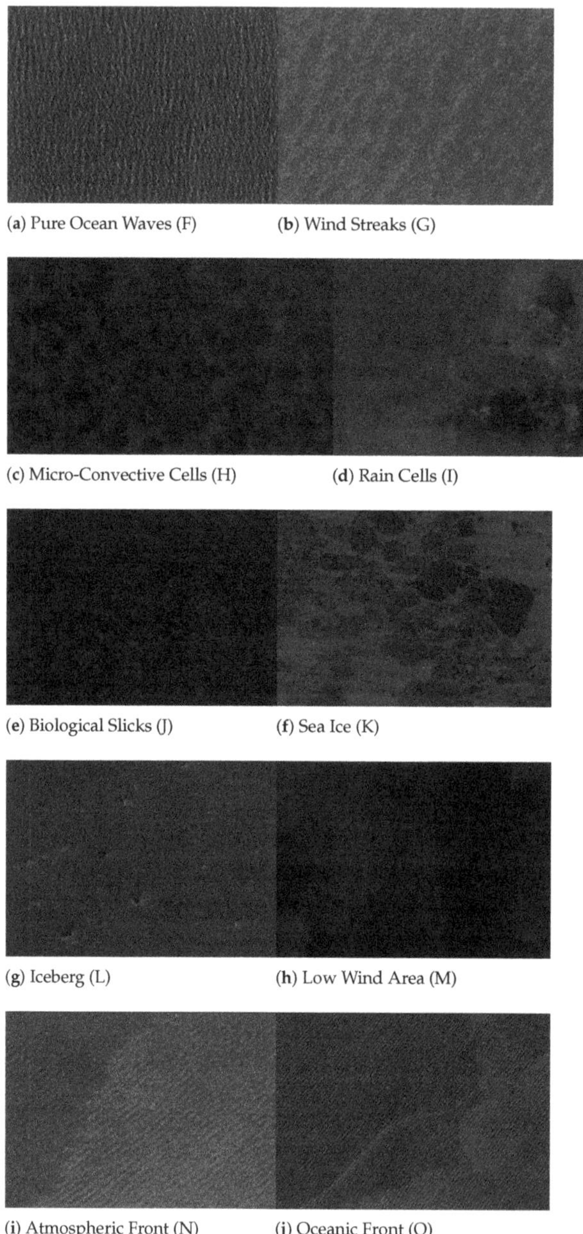

(a) Pure Ocean Waves (F) (b) Wind Streaks (G)

(c) Micro-Convective Cells (H) (d) Rain Cells (I)

(e) Biological Slicks (J) (f) Sea Ice (K)

(g) Iceberg (L) (h) Low Wind Area (M)

(i) Atmospheric Front (N) (j) Oceanic Front (O)

Figure 4. Dataset-2: Samples of SAR images from Sentinel-1 wave mode (TenGeoP-SARwv) dataset [59]. (**a**) Pure Ocean Waves, (**b**) Wind Streaks, (**c**) Micro-Convective Cells, (**d**) Rain Cells, (**e**) Biological Slicks, (**f**) Sea Ice, (**g**) Iceberg, (**h**) Low Wind Area, (**i**) Atmospheric Front, (**j**) Oceanic Front.

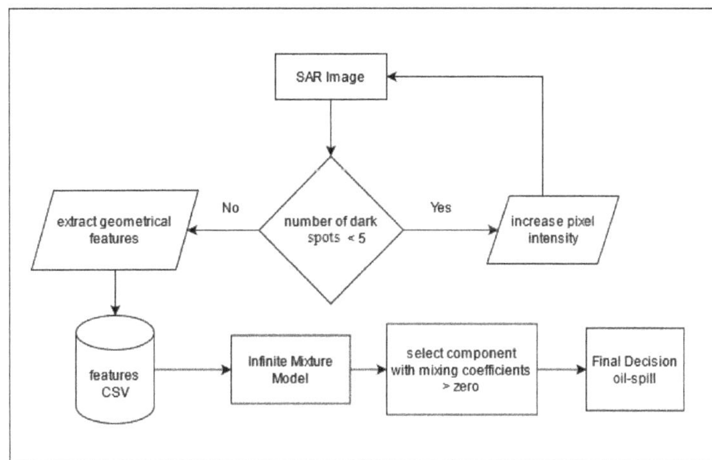

Figure 5. Flowchart diagram for extracting features using first feature extraction approach (ImageNet pretrained (resnet50) features).

Figures 6 and 7 present the confusion matrices for SAR images classification computed by the proposed InGaMM-eV using the two features extraction methods, respectively. It is noted that these matrices are used to describe the performance of the proposed model since they record true positives, false positives and false negatives. In fact, each matrix summarizes the prediction results on a classification problem and it offers a clear idea of what the proposed model is working correctly and what kinds of errors it commits. Each entry of index (u, v) represents the number of images in class u that are affected to class v. According to these results, the average classification accuracy is very promising and is equal to 90.57% (error rate of 9%) for the first dataset and 95.16% (error rate of 4%) for the second dataset.

Table 1. Results for both dataset with different number of classes using first feature extraction approach (ImageNet pretrained (resnet50) features).

Datasets	No of Class	Accuracy (%)	FPR
ESA-SAR dataset	2	97.96	0.02
ESA-SAR dataset	4	90.57	0.09
Sentinel-1 wave mode SAR dataset	2	94.53	0.05
Sentinel-1 wave mode SAR dataset	9	95.16	0.04

Table 2. Results for both dataset with different number of classes using second feature extraction approach (Dark spots, geometrical, physical, and characteristics features).

Datasets	No of Class	Accuracy (%)	FPR
ESA-SAR dataset	2	89.94	0.09
ESA-SAR dataset	4	85.13	0.12
Sentinel-1 wave mode SAR dataset	2	88.68	0.11
Sentinel-1 wave mode SAR dataset	9	82.22	0.14

Figures 6 and 7 present additional results obtained by changing the way visual features are extracted as well as the number of classes. Indeed, for the case of ESA-SAR dataset, InGaMM-eV provides high average accuracy of 97.96% using imageNet pretrained deep learning model (resnet50), and 89.94% using Dark spots, geometrical, physical characteristics features. In both cases, the false positive rate is very low. For Sentinel-1 wave mode SAR dataset, the average accuracy to classify SAR images is 95.16% using resnet50, which is better than the second method for extracting features (only 88.68%). According to these results, we notice that the overall average classification accuracy is very encouraging, taking into account the complexity of treated images. It is noteworthy that, due to low resolution of images in the second dataset (Sentinel-1 wave mode), it was very difficult to extract features using the second feature extraction method (i.e., detecting dark objects). Thus, we have low accuracy than expected for this dataset.

	Ships	Oil-spill	Land	Look-alike
Ships	87	0	2	11
Oil-spill	0	98	0	2
Land	0	0	96	4
Look-alike	0	0	14	86

Figure 6. Average rounded confusion matrix (in terms of percentage) for SAR classification using InGaMM-eV for ESA-SAR dataset.

	F	G	H	I	J	L	M	N	O
F	100	0	0	0	0	0	0	0	0
G	3	97	0	0	0	0	0	0	0
H	0	0	98	2	0	0	0	0	0
I	0	0	0	88	0	10	2	0	0
J	0	0	0	0	100	0	0	0	0
L	2	0	0	0	0	100	0	0	0
M	0	0	0	0	4	0	96	0	0
N	0	0	4	14	0	0	0	82	0
O	0	0	0	0	0	0	0	0	100

Figure 7. Average rounded confusion matrix (in terms of percentage) for SAR classification using InGaMM-eV for Sentinel-1 wave mode SAR dataset.

In this experiment, our second goal is also to demonstrate the advantages of using extended variational framework over the maximum likelihood (via EM-algorithm), as well as the merits of infinite mixture model over its finite counterpart. Therefore, we compared the classification results using the following mixture models: InGaMM-eV (our infinite Gamma model using extended variational inference), GaMM-eV (finite Gamma model using extended variational learning), GaMM-EM (finite Gamma mixture model using expectation maximization learning), InGMM-eV (infinite Gaussian model using extended variational learning), and GMM-EM (finite Gaussian mixture model using expectation maximization learning). The average performances of all tested learning approaches, using the two features extraction methods, are depicted in Tables 3 and 4. We can see clearly that the extended variational approach provides better results than the EM. Furthermore, the merits of using a Dirichlet process mixtures of Gamma distributions (i.e. infinite mixture model) over a finite mixture model is clear by noting that better result was found with the infinite mixtures. In particular, in Table 3, the InGaMM-eV (90.05%) outperforms GaMM-eV (88.33%) in terms of classification accuracy rate for both datasets. On the other side, it is worth mentioning that our approach provides better results than the implemented frameworks based on Gaussian mixtures. We can then deduce that the infinite Gamma model has better modeling and classification capability than the Gaussian when dealing with SAR images analysis.

Table 3. Overall oil spill detection rate of different models for 2 datasets using the first feature extraction approach (ImageNet pretrained (resnet50) features) .

Dataset	InGaMM-eV (*Our Approach*)	GaMM-eV	GaMM-EM	InGMM-eV	GMM-EM
ESA-SAR	90.05	88.33	86.07	83.21	83.11
Sentinel-1 wave SAR	91.12	89.40	87.02	84.14	83.99

Table 4. Overall oil spill detection rate of different models for 2 datasets using the second feature extraction approach (Dark spots, geometrical, physical, and characteristics features).

Dataset	InGaMM-eV (*Our Approach*)	GaMM-eV	GaMM-EM	InGMM-eV	GMM-EM
ESA-SAR	88.18	87.09	85.11	82.13	82.01
Sentinel-1 wave SAR	89.12	88.11	86.00	83.77	83.07

Next, The proposed learning approach (InGaMM-eV) is compared with some methods from the literature and the comparative study is presented in Table 5. As we can see, the proposed online algorithm performs better than other algorithms. Accordingly, it is important to emphasize the advantage of our developed extended variational formalism for infinite Gamma mixture, which can provide interesting results. It is also important to underline the merit of the online learning process, which is able to maintain high performance of oil spill prediction as well as handling data faster as they arrived. Moreover, it has the capacity to update the model incrementally without the need for retraining. All these results confirm that the proposed infinite Gamma mixture using the extended variational learning mode is a better choice thanks to the flexibility of the infinite Gamma mixture over the finite models. All these benefits make it more appropriate especially for SAR images classification especially in the case or large scale data sets.

Table 5. Comparative study between different methods from the literature on two datasets.

Method	Dataset	Feature Selection	Accuracy
InGaMM-eV (our approach)	ESA-SAR	ImageNet pretrained (resnet50)	**97.96**%
InGaMM-eV (our approach)	ESA-SAR	Dark spots, geometrical, physical features	**89.94**%
Fuzzy classification [62]	ESA-SAR	Georeference, Land masking, and Filtering	88%
InGaMM-eV (our approach)	Sentinel-1 SAR	ImageNet pretrained (resnet50)	**94.53**%
InGaMM-eV (our approach)	Sentinel-1 SAR	Dark spots, geometrical, physical features	**88.68**%
Convolutional neural network	Sentinel-1 SAR	Inception v3 CNN	93%
Articial neural network [34]	Sentinel-1 SAR	Dark spot, shape features	87%
Method in [63]	Sentinel-1 SAR	Dark spot features	81%
Method in [64]	Sentinel-1 SAR	Dark spot, shape features	82.61%

7. Conclusions

In this paper an effective online nonparametric Bayesian analysis method based on Dirichlet process mixture of Gamma distributions (i.e., infinite Gamma mixture model) is developed to deal with the challenging problem of oil spill detection in SAR images. The Gamma distribution is considered because of its flexibility for semi-bounded data modelling. This framework is learned using an extended version of conventional variational inference in a flexible way which has certain advantages such as approximating the posteriors effectively in a closed form, easy assessment of convergence and easy optimization by offering a trade-off between frequentist techniques and MCMC-based ones.

An important property of our approach is that it does not need the specification of the number of mixture components in advance. The proposed online algorithm has also the benefit to allow data instances to be treated in a sequential manner, which is more attractive than batch learning especially when dealing with massive and streaming data. Through the challenging application of oil spill detection in SAR images, we have demonstrated the performance of our statistical framework, which is able to provide very encouraging results in terms of SAR images modeling and classification capabilities. As future work, we plan to integrate a feature selection mechanism into the proposed framework in order to improve more the classification accuracy. It is our hope that many other real-world applications related to image processing and machine learning can be addressed via our developed framework.

Author Contributions: Conceptualization, A.A. and S.B.; methodology, N.B. and F.A.; software, Y.P.; validation, R.A., S.B. and A.A.; formal analysis, F.A.; investigation, A.A.; resources, Y.P.; data curation, S.B.; writing—original draft preparation, A.A. and S.B.; writing—review and editing, N.B. and F.A.; visualization, R.A.; supervision, N.B.; project administration, A.A.; funding acquisition, A.A. All authors have read and agreed to the published version of the manuscript.

Funding: This research was funded by Deanship of Scientific Research, Taif University, Kingdom of Saudi Arabia, grant number 1-441-137.

Institutional Review Board Statement: Not applicable.

Informed Consent Statement: Not applicable.

Data Availability Statement: Data available in a publicly accessible repository: The first data set: European Space Agency (ESA) database [40]. The second data set: Sentinel-1 wave mode (TenGeoP-SARwv) [59].

Conflicts of Interest: The authors declare no conflict of interest.

Appendix A

(1) Optimal solution to $Q(Z)$.

$$Q(Z) = \prod_{i=1}^{N}\prod_{j=1}^{M} r_{ij}^{Z_{ij}} \tag{A1}$$

where the responsibility r_{ij} can be calculated as:

$$r_{ij} = \frac{\rho_{ij}}{\sum_{j=1}^{M} \rho_{ij}} \tag{A2}$$

such that:

$$ln(\rho_{ij}) = ln(\pi_j) + \sum_{d=1}^{D} [\mathcal{P}_{jd} + (\langle\alpha_{jd}\rangle - 1) ln(y_{jd}) - \langle\alpha_{jd}\rangle\langle\beta_{jd}\rangle y_{jd}] \tag{A3}$$

and

$$\mathcal{P}_{jd} = \alpha_{jd}^{*} ln(\alpha_{jd}^{*}) - \alpha_{jd}^{*} - ln(\alpha_{jd}^{*}) - ln(\Gamma(\alpha_{jd}^{*})) + \langle ln(\alpha_{jd})\rangle + \langle\alpha_{jd}\rangle + \langle\alpha_{jd}\rangle\langle ln(\beta_{jd})\rangle \tag{A4}$$

where $\langle . \rangle$ refers to an expectation w.r.t. the corresponding factor and α_{jd}^{*} is any feasible point.

The expectation of Z_{ij} is determined as:

$$\langle Z_{ij}\rangle = r_{ij} \tag{A5}$$

(2) Optimal solution to $Q(\psi)$ and $Q(\lambda)$.

$$Q(\psi) = \prod_{j=1}^{M} \mathcal{G}(\psi_j \mid a_j, b_j) \tag{A6}$$

$$Q(\lambda) = \prod_{j=1}^{M} Beta(\lambda_j \mid c_j, d_j) \tag{A7}$$

$$\begin{aligned} a_j^* &= a_j + 1 \\ b_j^* &= b_j \ - \langle ln(1-\lambda_j) \rangle \\ c_j^* &= 1 + \sum_{i=1}^{N} \langle Z_{ij} \rangle \\ d_j^* &= \langle \psi_j \rangle + \sum_{i=1}^{N} \sum_{s=j+1}^{M} \langle Z_{is} \rangle \end{aligned} \tag{A8}$$

From the previous equations, we obtain the following expectations:

$$\begin{aligned} \langle ln(\lambda_j) \rangle &= \Psi(c_j^*) - \Psi(c_j^* + d_j^*) \\ \langle ln(1-\lambda_j) \rangle &= \Psi(d_j^*) - \Psi(c_j^* + d_j^*) \\ \langle ln(\psi_j) \rangle &= \frac{a_j^*}{b_j^*} \\ \langle \lambda_j \rangle &= \frac{c_j}{c_j + d_j} \end{aligned} \tag{A9}$$

where Ψ is Digamma function.

(3) Optimal solution to$Q(\vec{\alpha})$.

$$Q(\vec{\alpha}) = \prod_{j=1}^{M} \prod_{d=1}^{D} \mathcal{G}(\alpha_{jd} \mid u_{jd}^*, v_{jd}^*) \tag{A10}$$

where

$$\begin{aligned} u_{jd}^* &= u_{jd} + \sum_{i=1}^{N} \langle z_{ij} \rangle \\ v_{jd}^* &= v_{jd} - \sum_{i=1}^{N} [S_{jd} + ln(y_{id}) - \langle \beta_{jd} \rangle y_{id}] \langle z_{ij} \rangle \\ S_{jd} &= 1 + ln(\alpha_{jd}^*) - \frac{1}{\alpha_{jd}^*} - \Psi(\alpha_{jd}^*) + \langle ln(\beta_{jd}) \rangle \end{aligned} \tag{A11}$$

From the previous equations, we obtain the following expectations:

$$\begin{aligned} \langle \alpha_{jd} \rangle &= \frac{u_{jd}^*}{v_{jd}^*} \\ \langle ln(\alpha_{jd}) \rangle &= \Psi(u_{jd}^*) - ln(v_{jd}^*) \end{aligned} \tag{A12}$$

(4) Optimal solution to $Q(\vec{\beta})$.

$$Q(\vec{\beta}) = \prod_{j=1}^{M} \prod_{d=1}^{D} \mathcal{G}(\beta_{jd} \mid s_{jd}^*, t_{jd}^*) \tag{A13}$$

where

$$s_{jd}^* = s_{jd} + \langle \alpha_{jd} \rangle \sum_{i=1}^{N} \langle z_{ij} \rangle$$
$$t_{jd}^* = t_{jd} + \langle \alpha_{jd} \rangle \sum_{i=1}^{N} \langle z_{ij} \rangle y_{id} \quad \text{(A14)}$$

From the previous equations, we obtain the following expectations:

$$\langle \beta_{jd} \rangle = \frac{s_{jd}^*}{t_{jd}^*}$$
$$\langle ln(\beta_{jd}) \rangle = \Psi(s_{jd}^*) - ln(t_{jd}^*) \quad \text{(A15)}$$

References

1. Lary, D.J.; Alavi, A.H.; Gandomi, A.H.; Walker, A.L. Machine learning in geosciences and remote sensing. *Geosci. Front.* **2016**, *7*, 3–10. [CrossRef]
2. Lai, Y.; Ping, Y.; He, W.; Wang, B.; Wang, J.; Zhang, X. Variational Bayesian inference for finite inverted Dirichlet mixture model and its application to object detection. *Chin. J. Electron.* **2018**, *27*, 603–610. [CrossRef]
3. McLachlan, G.J.; Peel, D. *Finite Mixture Models*; John Wiley & Sons: Hoboken, NJ, USA, 2004.
4. Andrews, J.L.; McNicholas, P.D.; Subedi, S. Model-based classification via mixtures of multivariate t-distributions. *Comput. Stat. Data Anal.* **2011**, *55*, 520–529. [CrossRef]
5. Bouguila, N.; Almakadmeh, K.; Boutemedjet, S. A finite mixture model for simultaneous high-dimensional clustering, localized feature selection and outlier rejection. *Expert Syst. Appl.* **2012**, *39*, 6641–6656. [CrossRef]
6. Elguebaly, T.; Bouguila, N. Background subtraction using finite mixtures of asymmetric Gaussian distributions and shadow detection. *Mach. Vis. Appl.* **2014**, *25*, 1145–1162. [CrossRef]
7. Elguebaly, T.; Bouguila, N. Bayesian Learning of Generalized Gaussian Mixture Models on Biomedical Images. In *Artificial Neural Networks in Pattern Recognition, Proceedings of the 4th IAPR TC3 Workshop, ANNPR 2010, Cairo, Egypt, 11–13 April 2010*; Schwenker, F., Gayar, N.E., Eds.; Springer: Berlin, Germany, 2010; Volume 5998, pp. 207–218. [CrossRef]
8. Lai, Y.; Cao, H.; Luo, L.; Zhang, Y.; Bi, F.; Gui, X.; Ping, Y. Extended variational inference for gamma mixture model in positive vectors modeling. *Neurocomputing* **2021**, *432*, 145–158. [CrossRef]
9. Li, H.; Krylov, V.A.; Fan, P.; Zerubia, J.; Emery, W.J. Unsupervised Learning of Generalized Gamma Mixture Model With Application in Statistical Modeling of High-Resolution SAR Images. *IEEE Trans. Geosci. Remote Sens.* **2016**, *54*, 2153–2170. [CrossRef]
10. Ziou, D.; Bouguila, N. Unsupervised Learning of a Finite Gamma Mixture Using MML: Application to SAR Image Analysis. In Proceedings of the 17th International Conference on Pattern Recognition, (ICPR 2004), Cambridge, UK, 23–26 August 2004; pp. 68–71. [CrossRef]
11. Al-Osaimi, F.R.; Bouguila, N. A Finite Gamma Mixture Model-Based Discriminative Learning Frameworks. In Proceedings of the 14th IEEE International Conference on Machine Learning and Applications, ICMLA 2015, Miami, FL, USA, 9–11 December 2015; Li, T., Kurgan, L.A., Palade, V., Goebel, R., Holzinger, A., Verspoor, K., Wani, M.A., Eds.; IEEE: New York, NY, USA, 2015; pp. 819–824. [CrossRef]
12. Beckmann, C.; Woolrich, M.; Smith, S. Gaussian/Gamma mixture modelling of ICA/GLM spatial maps, In Proceedings of the 9th International Conference on Functional Mapping of the Human Brain. New York, NY, USA, 19–22 June 2003.
13. Alharithi, F.S.; Almulihi, A.H.; Bourouis, S.; Alroobaea, R.; Bouguila, N. Discriminative Learning Approach Based on Flexible Mixture Model for Medical Data Categorization and Recognition. *Sensors* **2021**, *21*, 2450. [CrossRef]
14. Bourouis, S.; Channoufi, I.; Alroobaea, R.; Rubaiee, S.; Andejany, M.; Bouguila, N. Color object segmentation and tracking using flexible statistical model and level-set. *Multim. Tools Appl.* **2021**, *80*, 5809–5831. [CrossRef]
15. Fan, W.; Bouguila, N.; Bourouis, S.; Laalaoui, Y. Entropy-based variational Bayes learning framework for data clustering. *IET Image Process.* **2018**, *12*, 1762–1772. [CrossRef]
16. Najar, F.; Bourouis, S.; Zaguia, A.; Bouguila, N.; Belghith, S. Unsupervised Human Action Categorization Using a Riemannian Averaged Fixed-Point Learning of Multivariate GGMM. In Proceedings of the Image Analysis and Recognition—15th International Conference, ICIAR 2018, Póvoa de Varzim, Portugal, 27–29 June 2018; pp. 408–415.
17. Ferguson, T.S. Bayesian density estimation by mixtures of normal distributions. In *Recent Advances in Statistics;* Academic Press: New York, NY, USA, 1983; pp. 287–302.
18. Rasmussen, C.E. A Practical Monte Carlo Implementation of Bayesian Learning. In Proceedings of the Advances in Neural Information Processing Systems 8, NIPS, Denver, CO, USA, 27–30 November 1995; Touretzky, D.S., Mozer, M., Hasselmo, M.E., Eds.; MIT Press: Cambridge, MA, USA, 1995; pp. 598–604.

19. Bourouis, S.; Alroobaea, R.; Rubaiee, S.; Andejany, M.; Almansour, F.M.; Bouguila, N. Markov Chain Monte Carlo-Based Bayesian Inference for Learning Finite and Infinite Inverted Beta-Liouville Mixture Models. *IEEE Access* **2021**, *9*, 71170–71183. [CrossRef]
20. Bouguila, N.; Elguebaly, T. A fully Bayesian model based on reversible jump MCMC and finite Beta mixtures for clustering. *Expert Syst. Appl.* **2012**, *39*, 5946–5959. [CrossRef]
21. Jordan, M.I.; Ghahramani, Z.; Jaakkola, T.S.; Saul, L.K. An Introduction to Variational Methods for Graphical Models. *Mach. Learn.* **1999**, *37*, 183–233. [CrossRef]
22. Fan, W.; Bouguila, N. Online Learning of a Dirichlet Process Mixture of Beta-Liouville Distributions Via Variational Inference. *IEEE Trans. Neural Netw. Learn. Syst.* **2013**, *24*, 1850–1862. [CrossRef] [PubMed]
23. Elguebaly, T.; Bouguila, N. A Bayesian approach for SAR images segmentation and changes detection. In Proceedings of the 2010 25th Biennial Symposium on Communications, Kingston, ON, Canada, 12–14 May 2010; pp. 24–27. [CrossRef]
24. Zhao, J.; Temimi, M.; Ghedira, H.; Hu, C. Exploring the potential of optical remote sensing for oil spill detection in shallow coastal waters-a case study in the Arabian Gulf. *Opt. Express* **2014**, *22*, 13755–13772. [CrossRef]
25. Singha, S.; Bellerby, T.J.; Trieschmann, O. Detection and classification of oil spill and look-alike spots from SAR imagery using an Artificial Neural Network. In Proceedings of the 2012 IEEE International Geoscience and Remote Sensing Symposium, IGARSS 2012, Munich, Germany, 22–27 July 2012; pp. 5630–5633.
26. Brekke, C.; Solberg, A.H. Oil spill detection by satellite remote sensing. *Remote Sens. Environ.* **2005**, *95*, 1–13. [CrossRef]
27. Salberg, A.; Larsen, S.O. Classification of Ocean Surface Slicks in Simulated Hybrid-Polarimetric SAR Data. *IEEE Trans. Geosci. Remote Sens.* **2018**, *56*, 7062–7073. [CrossRef]
28. Alpers, W.; Holt, B.; Zeng, K. Oil spill detection by imaging radars: Challenges and pitfalls. *Remote Sens. Environ.* **2017**, *201*, 133–147. [CrossRef]
29. Solberg, A.H.S.; Storvik, G.; Solberg, R.; Volden, E. Automatic detection of oil spills in ERS SAR images. *IEEE Trans. Geosci. Remote Sens.* **1999**, *37*, 1916–1924. [CrossRef]
30. Skrunes, S.; Brekke, C.; Eltoft, T. Characterization of Marine Surface Slicks by Radarsat-2 Multipolarization Features. *IEEE Trans. Geosci. Remote Sens.* **2014**, *52*, 5302–5319. [CrossRef]
31. Fingas, M.; Brown, C.E. A Review of Oil Spill Remote Sensing. *Sensors* **2018**, *18*, 91. [CrossRef]
32. Fiscella, B.; Giancaspro, A.; Nirchio, F.; Pavese, P.; Trivero, P. Oil spill detection using marine SAR images. *Int. J. Remote Sens.* **2000**, *21*, 3561–3566. [CrossRef]
33. Gambardella, A.; Giacinto, G.; Migliaccio, M.; Montali, A. One-class classification for oil spill detection. *Pattern Anal. Appl.* **2010**, *13*, 349–366. [CrossRef]
34. Topouzelis, K.; Karathanassi, V.; Pavlakis, P.; Rokos, D. Detection and discrimination between oil spills and look-alike phenomena through neural networks. *ISPRS J. Photogramm. Remote Sens.* **2007**, *62*, 264–270. [CrossRef]
35. Karantzalos, K.; Argialas, D. Automatic detection and tracking of oil spills in SAR imagery with level set segmentation. *Int. J. Remote Sens.* **2008**, *29*, 6281–6296. [CrossRef]
36. Chang, L.; Tang, Z.S.; Chang, S.H.; Chang, Y. A region-based GLRT detection of oil spills in SAR images. *Pattern Recognit. Lett.* **2008**, *29*, 1915–1923. [CrossRef]
37. Solberg, A.H.S.; Brekke, C.; Husoy, P.O. Oil Spill Detection in Radarsat and Envisat SAR Images. *IEEE Trans. Geosci. Remote Sens.* **2007**, *45*, 746–755. [CrossRef]
38. Keramitsoglou, I.; Cartalis, C.; Kiranoudis, C.T. Automatic identification of oil spills on satellite images. *Environ. Model. Softw.* **2006**, *21*, 640–652. [CrossRef]
39. Cantorna, D.; Dafonte, C.; Iglesias, A.; Varela, B.A. Oil spill segmentation in SAR images using convolutional neural networks. A comparative analysis with clustering and logistic regression algorithms. *Appl. Soft Comput.* **2019**, *84*, 105716. [CrossRef]
40. Krestenitis, M.; Orfanidis, G.; Ioannidis, K.; Avgerinakis, K.; Vrochidis, S.; Kompatsiaris, I. Oil Spill Identification from Satellite Images Using Deep Neural Networks. *Remote Sens.* **2019**, *11*, 1762. [CrossRef]
41. Orfanidis, G.; Ioannidis, K.; Avgerinakis, K.; Vrochidis, S.; Kompatsiaris, I. A Deep Neural Network for Oil Spill Semantic Segmentation in Sar Images. In Proceedings of the 2018 IEEE International Conference on Image Processing, ICIP 2018, Athens, Greece, 7–10 October 2018; pp. 3773–3777.
42. Song, D.; Ding, Y.; Li, X.; Zhang, B.; Xu, M. Ocean Oil Spill Classification with RADARSAT-2 SAR Based on an Optimized Wavelet Neural Network. *Remote Sens.* **2017**, *9*, 799. [CrossRef]
43. Li, J.; Du, Q.; Li, Y. An efficient radial basis function neural network for hyperspectral remote sensing image classification. *Soft Comput.* **2016**, *20*, 4753–4759. [CrossRef]
44. Shaban, M.; Salim, R.; Abu Khalifeh, H.; Khelifi, A.; Shalaby, A.; El-Mashad, S.; Mahmoud, A.; Ghazal, M.; El-Baz, A. A Deep-Learning Framework for the Detection of Oil Spills from SAR Data. *Sensors* **2021**, *21*, 2351. [CrossRef]
45. Zeng, K.; Wang, Y. A Deep Convolutional Neural Network for Oil Spill Detection from Spaceborne SAR Images. *Remote Sens.* **2020**, *12*, 1015. [CrossRef]
46. Topouzelis, K.N. Oil Spill Detection by SAR Images: Dark Formation Detection, Feature Extraction and Classification Algorithms. *Sensors* **2008**, *8*, 6642–6659. [CrossRef]
47. Teh, Y.W. Dirichlet Process. In *Encyclopedia of Machine Learning*; Sammut, C., Webb, G.I., Eds.; Springer: Boston, MA, USA, 2010; pp. 280–287.
48. Sethuraman, J. A constructive definition of Dirichlet priors. *Stat. Sin.* **1994**, *4*, 639–650.

49. Blei, D.M.; Jordan, M.I. Variational inference for Dirichlet process mixtures. *Bayesian Anal.* **2006**, *1*, 121–143. [CrossRef]
50. Ishwaran, H.; James, L.F. Gibbs sampling methods for stick-breaking priors. *J. Am. Stat. Assoc.* **2001**, *96*, 161–173. [CrossRef]
51. Opper, M.; Saad, D. *Advanced Mean Field Methods: Theory and Practice*; MIT Press: Cambridge, MA, USA, 2001.
52. Blei, D.M.; Jordan, M.I. Variational methods for the Dirichlet process. In *Machine Learning, Proceedings of the Twenty-First International Conference (ICML 2004), Banff, AL, Canada, 4–8 July 2004*; Brodley, C.E., Ed.; ACM International Conference Proceeding Series; ACM: New York, NY, USA, 2004; Volume 69.
53. Sato, M. Online Model Selection Based on the Variational Bayes. *Neural Comput.* **2001**, *13*, 1649–1681. [CrossRef]
54. Fan, W.; Bouguila, N. Online variational learning of generalized Dirichlet mixture models with feature selection. *Neurocomputing* **2014**, *126*, 166–179. [CrossRef]
55. Hoffman, M.D.; Blei, D.M.; Bach, F.R. Online Learning for Latent Dirichlet Allocation. In *Advances in Neural Information Processing Systems*; Curran Associates, Inc.: New York, NY, USA, 2010; pp. 856–864.
56. Manouchehri, N.; Nguyen, H.; Koochemeshkian, P.; Bouguila, N.; Fan, W. Online Variational Learning of Dirichlet Process Mixtures of Scaled Dirichlet Distributions. *Inf. Syst. Front.* **2020**, *22*, 1085–1093. [CrossRef]
57. Konik, M.; Bradtke, K. Object-oriented approach to oil spill detection using ENVISAT ASAR images. *ISPRS J. Photogramm. Remote Sens.* **2016**, *118*, 37–52. [CrossRef]
58. Topouzelis, K.; Psyllos, A. Oil spill feature selection and classification using decision tree forest on SAR image data. *ISPRS J. Photogramm. Remote Sens.* **2012**, *68*, 135–143. [CrossRef]
59. Wang, C.; Mouche, A.; Tandeo, P.; Stopa, J.E.; Longépé, N.; Erhard, G.; Foster, R.C.; Vandemark, D.; Chapron, B. A labelled ocean SAR imagery dataset of ten geophysical phenomena from Sentinel-1 wave mode. *Geosci. Data J.* **2019**, *6*, 105–115. [CrossRef]
60. He, K.; Zhang, X.; Ren, S.; Sun, J. Deep Residual Learning for Image Recognition. In Proceedings of the 2016 IEEE Conference on Computer Vision and Pattern Recognition, CVPR 2016, Las Vegas, NV, USA, 27–30 June 2016; IEEE Computer Society: Washington, DC, USA, 2016; pp. 770–778.
61. Topouzelis, K.; Stathakis, D.; Karathanassi, V. Investigation of genetic algorithms contribution to feature selection for oil spill detection. *Int. J. Remote Sens.* **2009**, *30*, 611–625. [CrossRef]
62. Ferraro, G.; Pavlakis, P.; Tarchi, D.; Sieber, A.; Ferraro, G.; Vincent, G. *On the Monitoring of Illicit Discharges—A Reconnaissance Study in the Mediterranean Sea*; EUR 19906 EN; European Commission: Brussels, Belgium, 2001.
63. Chatziantoniou, A.; Karagaitanakis, A.; Bakopoulos, V.; Papandroulakis, N.; Topouzelis, K. Detection of Biogenic Oil Films near Aquaculture Sites Using Sentinel-1 and Sentinel-2 Satellite Images. *Remote Sens.* **2021**, *13*, 1737. [CrossRef]
64. Chatziantoniou, A.; Bakopoulos, V.; Papandroulakis, N.; Topouzelis, K. Detection of biogenic oil film near aquaculture sites seen by Sentinel-2 multispectral images. In *Remote Sensing of the Ocean, Sea Ice, Coastal Waters, and Large Water Regions 2020*; International Society for Optics and Photonics: San Diego, CA, USA, 2020; p. 4. [CrossRef]

Article

High Wind Speed Inversion Model of CYGNSS Sea Surface Data Based on Machine Learning

Yun Zhang [1], Jiwei Yin [1], Shuhu Yang [1,*], Wanting Meng [2], Yanling Han [1] and Ziyu Yan [1]

1 Shanghai Marine Intelligent Information and Navigation Remote Sensing Engineering Technology Research Center, Shanghai Ocean University, Shanghai 201306, China; y-zhang@shou.edu.cn (Y.Z.); m190711309@st.shou.edu.cn (J.Y.); ylhan@shou.edu.cn (Y.H.); m200711477@st.shou.edu.cn (Z.Y.)

2 Shanghai Spaceflight Institute of TT&C and Telecommunication, Shanghai 201109, China; wanting_meng@163.com

* Correspondence: shyang@shou.edu.cn

Abstract: In response to the deficiency of the detection capability of traditional remote sensing means (scatterometer, microwave radiometer, etc.) for high wind speed above 25 m/s, this paper proposes a GNSS-R technique combined with a machine learning method to invert high wind speed at sea surface. The L1-level satellite-based data from the Cyclone Global Navigation Satellite System (CYGNSS), together with the European Centre for Medium-Range Weather Forecasts (ECMWF) and the National Centers for Environmental Prediction (NCEP) data, constitute the original sample set, which is processed and trained with Support Vector Regression (SVR), the combination of Principal Component Analysis (PCA) and SVR (PCA-SVR), and Convolutional Neural Network (CNN) methods, respectively, to finally construct a sea surface high wind speed inversion model. The three models for high wind speed inversion are certified by the test data collected during Typhoon Bavi in 2020. The results show that all three machine learning models can be used for high wind speed inversion on sea surface, among which the CNN method has the highest inversion accuracy with a mean absolute error of 2.71 m/s and a root mean square error of 3.80 m/s. The experimental results largely meet the operational requirements for high wind speed inversion accuracy.

Keywords: GNSS-R; CYGNSS; high wind speed inversion; SVR; PCA-SVR; CNN

Citation: Zhang, Y.; Yin, J.; Yang, S.; Meng, W.; Han, Y.; Yan, Z. High Wind Speed Inversion Model of CYGNSS Sea Surface Data Based on Machine Learning. *Remote Sens.* **2021**, *13*, 3324. https://doi.org/10.3390/rs13163324

Academic Editors: Monidipa Das, Soumya K. Ghosh, V. M. Chowdary, Pabitra Mitra and Santosh Rijal

Received: 29 June 2021
Accepted: 19 August 2021
Published: 23 August 2021

Publisher's Note: MDPI stays neutral with regard to jurisdictional claims in published maps and institutional affiliations.

1. Introduction

As one of the most serious natural disasters in the world, typhoons are a top priority for scientific research because of their suddenness and destructive power, which bring huge economic losses to human society. Remote sensing technology provides a huge development space for typhoon monitoring and prediction. All microwave remote sensing instruments are struggling to provide reliable high wind speed measurements above 25 m/s. However, few studies have been obtained up to now [1–5]. The Global Navigation Satellite System reflection (GNSS-R) technology uses satellite signals reflected from the Earth's surface to obtain information of surface characteristics such as sea surface wind speed, so it can be provided with all-weather detection capability [6–11]. The main purpose of the Cyclone Global Navigation Satellite System (CYGNSS), launched by the United States in 2016, is to monitor tropical cyclones. It measures sea surface winds in and near the eyewalls of tropical cyclones, typhoons, and hurricanes frequently throughout their life cycle and the data collected can be used to invert wind speeds [12].

Many methods can be used to inverse wind speed. For example, a GNSS-R wind speed inversion method is to extract DDM observables reflecting the wind speed from the delay-Doppler map (DDM) and then build the Geophysical Model Function (GMF) model for wind speed inversion. Some other studies use the matched filter method between simulated DDMs and measured DDMs to inverse wind speed. In addition, the machine learning method is also suitable for wind speed inversion.

In 2014, Clarizia et al. [6] extracted five DDM observables from the United Kingdom disaster monitor constellation (UK-DMC) satellite, namely Delay-Doppler Map Average (DDMA), Delay-Doppler Map Variance (DDMV), Allan DDM Variance (ADDMV), Leading Edge Slope (LES), and Trailing Edge Slope (TES). Different GMF models were established by comparing the five observables with the buoy wind speed provided by the National Data Buoy Center (NDBC). Then, the reverse wind speeds, using each of these GMFs, were combined into a minimum variance estimator. The root mean square error (RMSE) obtained for wind speeds less than 10 m/s was 1.65 m/s. In 2016, Rodriguez et al. [13] used a generalized observable to determine the coefficients of linear combination by the maximum signal-to-noise ratio (MSNR), the minimum variance of the wind speed (MVU), and principal component analysis (PCA). Then, the three wind speed values were compared with the CYGNSS baseline L2 observables. The results show that PCA performs best, but the overall RMS was greater than 4 m/s when the wind speed was greater than 20 m/s. In 2018, Christopher S. Ruf et al. [9] used the observable DDMA and LES from CYGNSS L1 level data to establish the Fully Developed Seas (FDS) GMF and Young Seas/Limited Fetch (YSLF) GMF for different incidence angles at different seas. The knowledge of wind speed inversion algorithm required to establish FDS GMF model comes from [7]. The reference wind speed used to train the FDS GMF were the 10 m-referenced ocean surface wind speeds provided by the ECMWF and the GDAS. The YSLF GMF model was established using the wind speed collected by the stepped frequency microwave radiometer (SFMR) on NOAA P-3 hurricane hunter aircraft. The FDS GMF model was suitable for low-to-moderate wind speeds. On the contrary, the YSLF GMF model was more sensitive to hurricanes. By using FDS GMF to invert wind speed below 20 m/s and comparing with the European Centre for Medium-Range Weather Forecasts (ECMWF), the overall RMSE was about 2 m/s. In addition, compared with SFMR aircraft data, when the wind speed was greater than 20 m/s, the RMSE of the YSLF GMF model inversion wind speed was about 6.5 m/s. In addition, the samples for wind speeds greater than 20 m/s tested numbered only 674.

In 2017, F.Said [14] et al. proposed a method to inverse the maximum hurricane wind speed using the simulated power-versus-delay waveform data of CYGNSS. The CYGNSS end-to-end simulator (E2ES) [7] was used to produce the reference simulated waveforms. The specific process was to compare the simulated waveform with the reference waveform generated over a set of synthetic Willoughby storms with known maximum wind speed (Vmax) through the matched filter, and output the Vmax corresponding to the reference waveform. The Vmax was the retrieved wind speed. Comparing the retrieved Vmax values of 552 hurricane events with the hurricane weather research and forecasting model (HWRF) Vmax and the Best Track for Vmax, the overall bias of wind speed less than 40 m/s was greater than 11 m/s, and the overall bias of wind speed greater than 40 m/s was less than 3 m/s. However, the samples of hurricane wind speed studied were not enough. In 2019, Al-Khaldi, M [15] et al. extended the simulation study of [14] to the use of CYGNSS full DDM. A matched filter approach between normalized simulated DDMs and measured DDMs was applied to inverse storm parameters. The Vmax estimates were inversed by using the data during Hurricane Irma. Compared with the reported National Hurricane Center Best Track forecasts, the RMSE was 6.89 m/s. In 2021, the same team including Al-Khaldi, M [16] carried out a progress update and error analysis on the research performed by [15]. They continued to use the CYGNSS full DDM and proposed to use the synthetic storm model to retrieve wind speed on the basis of [15]. The synthetic storm model included the Willoughby model and Generalized Asymmetric Holland Model (GAHM). The success of inversion was due to the combination of the GAHM model suitable for storms with low levels of development and the Willoughby model suitable for storms with higher levels. The inversion Vmax was obtained by combining the results of the two models. Compared with the Best Track forecasts, the RMSE was 6.05 m/s. The RMSE was partially improved by comparison with the reference [15]. The effects of measurement delay extent on inverse error were also analyzed.

In 2019, Chong Wu et al. [17] used a back propagation (BP) neural network to invert the wind speed from 0 to 30 m/s, based mainly on the DDM data from CYGNSS. The DDM Observables included DDMA, LES, and Bistatic Radar Cross Section (BRCS). The paper used the CYGNSS L2 wind speed data as the reference wind speed. The Pearson correlation coefficient of the inverse wind speed and the CYGNSS wind speed data product was 0.958, the RMSE was 1.86 m/s, and the mean relative error was 2.66%. The feasibility and effectiveness of wind speed inversion using neural network based on DDM was demonstrated. However, the amount of data for wind speeds greater than 20 m/s in the paper was small and the applicability of the neural network for high wind speed data cannot be confirmed. In the same year, Han Gao et al. [18] used eight observables in CYGNSS L1 data (DDMA, LES, TES, specular reflection point position, satellite altitude angle, Scatter Area, delay-Doppler correlation power mean, and Effective Area) to train the model with a BP neural network, and then compared the reverse wind speed with the wind speed data provided by ECMWF. When the wind speed was less than 20 m/s, the RMSE was 1.21 m/s, and the RMSE in the wind speed range of 20~45 m/s was 2.54 m/s. However, this paper only had 4761 wind speed data above 20 m/s, which was not enough for high wind speed training.

In 2020, Jennifer et al. [10] proposed the Artificial Neural Network (ANN) inversion algorithm for wind speed inversion based on CYGNSS satellite data. In this paper, six characteristic parameters (DDMA, LES, Incidence Angle, Range Corrected Gain (RCG) [7], and Latitude and Longitude of the specular point acquisition.) were used to train ANN model, and CYGNSS L2 wind speed data was used as the reference wind speed. The RMSD of wind speed inversion error for the range of 0~32 m/s was 1.51 m/s. However, the wind speeds in the paper mainly focus on 0–20 m/s, and there was not enough research on wind speeds above 20 m/s, thus good inversion results cannot be obtained for tropical storms. In the same year, Sja Wang [19] performed a comparison between neural network and machine learning methods using Tech Demo Sat-1 (TDS-1) satellite DDM map data and ECMWF data for wind speeds in the 3–18 m/s interval. It was verified that the inversion effect of the neural network model had a significant advantage with a 20% performance improvement.

In 2020, Cardellach et al. [20] combined CYGNSS uncalibrated Level-1 bin original observation count with ECMWF/C3S ERA5 reanalysis dataset to obtain specular reflection point wind speed. The study covered hurricane season data for 2018 and 2019. The inversion was carried out by a variational technique based on physical forward model. The inverse wind speed was compared with the background model, other spaceborne sensors, such as NASA Soil Moisture Active Passive (SMAP), ESA Soil Moisture and Ocean Salinity (SMOS), EUMETSAT Advanced Scatterometer on board METOP (ASCAT) A/B, and other organizations' CYGNSS inverse wind speed. The research showed that this method had the ability to infer wind speed (including hurricane winds). The inverse wind speed was the most consistent with NOAA inversion [21], but the lowest correlation was found between inversion and the official products that were obtained with the YSLF GMF, and the dispersion reached 9.9 m/s. The author expected that this method will work at moderate wind speed, but this method had the possibility of underestimating wind speed.

According to the above research results, it can be found that machine learning has been widely used in the inversion of sea surface wind speed in the field of remote sensing at present; however, relevant studies for high wind speed greater than 20 m/s are relatively lacking [22].

In this paper, we put forward a high wind speed inversion model for CYGNSS data based on machine learning methods for inversion of typhoons. The datasets consist of the CYGNSS measured L1 data published by the National Aeronautics and Space Administration (NASA) and the reanalyzed wind speed datasets of the ECMWF and National Centers for Environmental Prediction (NCEP). Three methods, Support Vector Regression (SVR), the combination of PCA and SVR (PCA-SVR), and Convolutional Neural Networks (CNN), are used to train the wind speed data above 20 m/s. Due to the uneven distribution of samples, the under-sampling method is used to extract data for training. The three models obtained after training are used to inverse the high wind speed during the typhoon Bavi life cycle typhoon in 2020. Compared with the wind speed from ECMWF/NCEP, the inversion results are used to study the performance of the three models.

2. Materials and Methods

2.1. Data Source

2.1.1. CYGNSS

The CYGNSS satellites are a constellation of eight low Earth orbit (LEO) microsatellites launched in 2016. Each satellite is equipped with a right-hand-circular polarization (RHCP) antenna to receive direct signals from the transmitting satellite and two left-hand-circular polarization (LHCP) antennas to receive reflected signals from reflective surfaces such as the sea surface. The specular reflection points collected by the CYGNSS satellite cover approximately $\pm 40^{\circ}$ latitude zone in the global area, and the longitude zone is completely covered. CYGNSS seeks to improve weather prediction capabilities by studying the interaction between ocean surface properties, humid atmospheric thermodynamics, radiation, and convective dynamics associated with tropical cyclones [7,9,12,14–16,20]. CYGNSS data is encapsulated by NASA in netCDF file format, and this paper used version 2.1 of the CYGNSS Level 1 data (available online at https://podaac.jpl.nasa.gov/dataset/CYGNSS_L1_V2.1, accessed on 8 April 2021), which is the result of the power expression transformed by L0 level DDM [10].

2.1.2. Mean Sea Level Pressure

Mean sea level (MSL) pressure is an important factor affecting typhoon status and its path [23]. This paper uses the MSL pressure reanalysis data product provided by ECMWF's official website. The MSL pressure reanalysis dataset calculates the atmospheric pressure on the Earth's surface, including all land, ocean, and inland water, and then adjusts the surface atmospheric pressure height to the height of mean sea level. The spatial resolution of MSL pressure dataset is 0.5°, and the temporal resolution is 1 h.

2.1.3. Global Wind Speed Data

This paper used two different global reanalysis wind speed datasets: ECMWF reanalysis dataset and NCEP reanalysis dataset, mainly to study wind speed data at the 10 m-referenced ocean surface wind speed (u10), using UTC time. ECMWF regularly uses its forecasting models and data assimilation system to reanalyze archived observations and further create global reanalysis datasets describing the recent history of the atmosphere, land, and ocean. The datasets provide sea surface wind speed at a spatial and temporal resolution of 1 h, 0.5°. NCEP adopts a state-of-the-art global data assimilation system and a comprehensive database to quality control and assimilate observations from various sources (ground, ships, radio soundings, wind balloons, aircraft, satellites, etc.) to obtain reanalysis datasets. The datasets provide sea surface wind speed with a temporal and spatial resolution of 1 h, 0.2°. Further using the time, latitude, and longitude of the observed data provided by CYGNSS, the reanalysis datasets are passed through spatial linear interpolation with temporal linear interpolation to obtain the corresponding wind speed in time and space. This paper combined the wind speed reanalysis datasets of ECMWF and NCEP. The data from ECMWF alone were used when the wind speed is less than 20 m/s, and the data from NCEP are used when the wind speed was greater than 20 m/s [9,24,25]. Finally, the wind speed dataset was composed into new datasets according to this criterion, and the new datasets were used as the true wind speed for training and testing.

2.2. Machine Learning Methods

Three methods, SVR, PCA-SVR, and CNN, were used to train the data to obtain three models; the following sections briefly outline the principles of each method.

2.2.1. SVR

SVR can improve the generalization ability of model by seeking structural risk minimization, so as to achieve the minimum empirical risk and confidence interval. Using fewer samples can also obtain good statistical rules. The input data is normalized before the SVR training to prevent training imbalance caused by feature anomalies. Additionally, normalization can also improve the computational speed. The SVR algorithm first symmetrically maps the input data X into a multidimensional space in a nonlinear way and then performs linear programming in that space. The selection of parameters of SVR generally includes three elements: The first is the selection of kernel function, here the radial basis function (RBF) with better smoothing performance is chosen; the second parameter is the selection of penalty factor C; the third parameter is the selection of kernel coefficient gamma value. In order to avoid overfitting and underfitting, this paper uses the grid search method to perform parameter search for C and gamma values when training the model [26,27]. In order to improve the rate of parameter search, the grid search method is adjusted as follows. Firstly, by finding the optimal parameters in a wide range roughly, and then by setting a smaller step size to search again according to this optimal parameter taking range.

The goal of SVR can be formalized as:

$$
\begin{gathered}
\mathrm{Min}\frac{1}{2}\|\omega\|^2 + C\sum_{i=1}^{n}(\xi_i + \xi_i{}^*) \\
\text{s.t. } y_i - \omega\phi(x) - b \le \varepsilon + \xi_i \; \xi_i \ge 0 \\
\omega\phi(x) + b - y_i \le \varepsilon + \xi_i{}^* \; \xi_i{}^* \ge 0 \\
i = 1, 2, \ldots, n
\end{gathered}
\tag{1}
$$

where ω is the normal vector, which determines the direction of the hyperplane. n is the number of samples, $C > 0$ is the penalty parameter, ε is the error sensitivity index, and ξ_i and $\xi_i{}^*$ are slack variables. By using the dual principle and introducing Lagrange multipliers, the above formula is solved:

$$
f(x) = \sum_{i=1}^{n}(\beta_i{}^* - \beta_i)K(x_i, x_j) + b \tag{2}
$$

where $\beta_i{}^*$ and β_i are the Lagrange multipliers, $K(x_i, x_j)$ is the radial basis function, and b is the threshold. Equation (2) is a kernel function introduced by the nonlinear SVR to deal with dimensional catastrophes [28].

The preprocessed training data were trained by SVR method, and the gamma value of the model was determined to be 72.50 and C was 0.09 by grid search.

2.2.2. PCA-SVR

Since the number of features tends to increase the model training time, PCA was used here to reduce the dimensionality of SVR input by secondary integration of multidimensional feature covariates in order to reduce the model training time and improve the independence of feature covariates. PCA, as a technique of data dimensionality reduction, can project the original features to the dimension with the maximum amount of projected information as much as possible and ensure the minimum loss of information after dimensionality reduction without affecting the final model prediction results, the processed data are then fed into the SVR for data prediction [29,30].

In the PCA-SVR prediction model, a total of 27 influencing factors were used as input data in this paper. The input training set was processed by PCA to obtain the principal components PC1, PC2, … , and PCk ($k \le 27$) for model prediction, and it was found that the cumulative contribution of the first 13 principal components reached more than 85%, which could replace all feature covariates for model training, so k was 13. Then, the dimensionality reduction data was input into SVR, and the gamma value of the model was determined to be 32.50 and C was 0.37 using the grid search method. Figure 1 shows the structure of the PCA-SVR model.

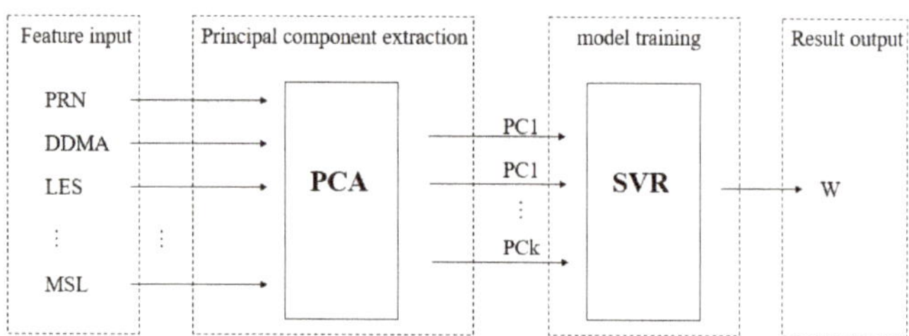

Figure 1. PCA-SVR model structure.

When the wind speed modeling is completed and enters the wind speed inversion stage, the feature parameters of the CYGNSS test set are normalized and directly multiplied with the corresponding feature vectors to obtain the principal component parameters. Then, the trained model is used for high wind speed inversion and the inversion accuracy of the inverse wind speed is calculated.

2.2.3. CNN

A CNN is a feed-forward neural network that performs well on image, audio, and text data. It is easy to update the data model by a back propagation algorithm. The CNN architecture (i.e., the number of layers and their structure) can be applied to a wide range of problems, while the hidden layers also reduce the algorithm's reliance on feature engineering. A CNN is suitable for training with large amounts of data and is capable of solving complex nonlinear problems. The complete neural network structure includes input layer, convolution layer, Relu activation function, pooling layer, fully connected layer, and output layer [19,31]. The optimizer uses adaptive moment estimation (Adam) gradient descent algorithm instead of stochastic gradient descent (SGD) because Adam is able to adjust the learning rate of each parameter, making the parameters smooth for extracting data features. A total of Xn samples are trained and the inversed wind speed values W are output.

After a large amount of data validation, this paper finally determined the number of convolutional layers to be 3, no pooling layer was set, the convolutional kernel size was 3 × 1, dropout was 0.3, the number of convolutional kernels in each layer was 32, batch-size was 1000, and epochs were 2000. Figure 2 shows the structure of the CNN model used in this paper.

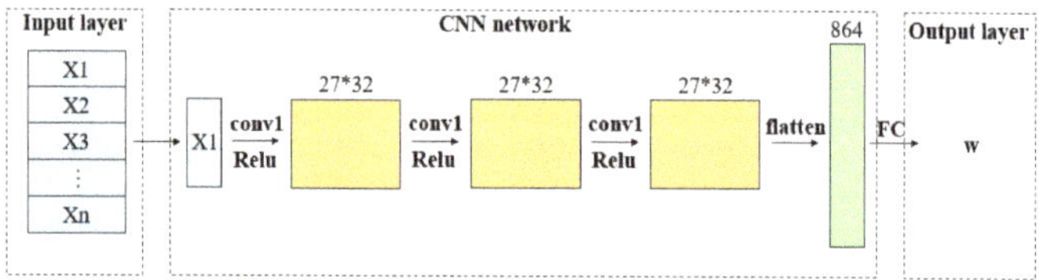

Figure 2. CNN model structure (Conv: Convolution layer, FC: fully connected layer).

2.3. High Wind Speed Inversion Process

2.3.1. Data Processing Flow

The process of high wind speed inversion in this paper can be briefly summarized into four parts: (i) determining the satellite data as well as wind speed data used; (ii) preprocessing and normalizing data; (iii) training the processed data with the three machine learning methods described above; (iv) using test data to inverse wind speed and analyzing performance of inversion wind speed. The specific wind speed inversion process is shown in Figure 3.

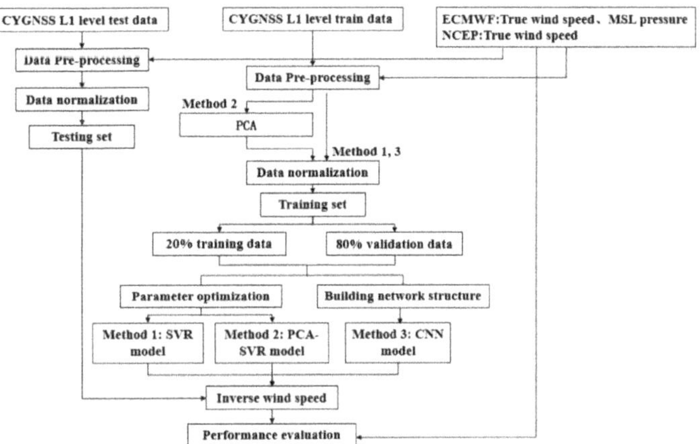

Figure 3. High wind speed inversion process based on machine learning.

2.3.2. Data Pre-Processing

In order to obtain good results, any data derived from a remote sensing satellite for Earth observation needs to undergo rigorous data pre-processing. The datasets were processed according to the following criteria:

(1) CYGNSS data quality control (QC) flags.
(2) Positive values for both CYGNSS observations and wind speed matching data.
(3) The RCG of the observations is greater than 10, with the RCG defined and described in [7].
(4) The incidence angle of the satellite antenna is less than 60°.
(5) The specular reflection point is at sea.

Because the occurrence time of each typhoon was not continuous, the CYGNSS data used in this paper was intermittent in time. CYGNSS data from 30 June 2018 to 3 July, 27 September 2018 to 30 September, 1 January 2019 to 3 January, 3 August 2019 to 8 August, 6 October 2019 to 12 October, 24 October 2019 to 25 October, 30 October 2019, 4 November 2019 to 7 November, 2 August 2020 to 4 August, 10 August 2020 to 11 August and 3 September 2020 were used as the train data (Figure 3), and the reanalysis wind speed datasets of ECMWF and NCEP in the corresponding time were used as the true wind speed data (Figure 3). Figure 4a,b respectively show the number of original samples and the corresponding number of final training samples for each wind speed range. The number of original samples means the data number after filtering the data according to the preprocessing criteria, and the number of final training samples means the training set data (including training data and validation data) number for Machine Leaning after under-sampling original samples.

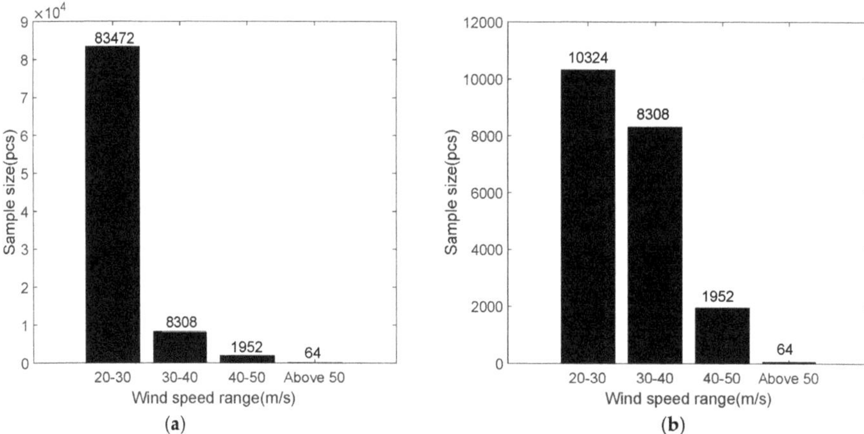

Figure 4. (**a**) Original training samples histogram; (**b**) Final training samples histogram.

From Figure 4a, we can see that the wind speed samples were concentrated between 20~30 m/s, the number of wind speed samples larger than 30 m/s and 20~30 m/s was seriously unbalanced, the imbalance of the number of samples easily led to the bias of the trained model, which did not have generalization. Therefore, the under-sampling method was used for random sampling to remove some majority samples from the training set, and in order to ensure that there were enough samples for training and that the amount of data for each type of wind speed interval was similar. Finally, when the ratio of samples between 20~30 m/s interval and more than 30 m/s interval was 1:1, a total of 20,648 final training samples were used for training. The specific samples are shown in Figure 4b. Subsequent model training and data research were based on this basis.

2.3.3. Feature Parameter Selection

After data pre-processing, it could be found that the L1 level data products of CYGNSS included many satellite observables, such as DDMA, LES, etc., which are characteristic values depending on wind speed as well as sea surface roughness. Due to the high wind speed measurement environment, especially typhoons, the sensitivity of the characteristic parameters of the two-dimensional delay-Doppler power waveform of the GNSS reflection signal to wind speed decreases, causing an increase in the wind speed measurement error. To reduce the performance error of CYGNSS in detecting typhoons, more characteristic parameters of CYGNSS L1 datasets were extracted to optimize the accuracy of the wind measurement model.

In this paper, 27 eigenvalues related to sea surface wind speed were used, specifically: Pseudo Random Noise (PRN) satellite number, DDMA, LES, antenna gain, distance from transmitter to specular reflection point, distance from receiver to specular reflection point, specular reflection point (longitude, latitude, time, and elevation angle), QC Flag, Signal-to-Noise Ratio (SNR), GNSS-R satellite position in ECEF, GNSS satellite position in ECEF, BRCS's DDM (specular delay line and Doppler column), BRCS's DDM (peak delay line and peak Doppler column), vehicle's specular delay, corrected DDM instrument specular delay, the direct signal code phase, and MSL pressure.

3. Results and Discussion

3.1. Typhoon Validation Data

To analyze the feasibility of the three methods for wind speed inversion, the data of the Typhoon Bavi event in August 2020 were studied here. The reflected signal data collected by CYGNSS during Typhoon Bavi in the western Pacific Ocean during 2020.8.22~2020.8.26 were processed as test data (Figure 3). The reanalysis typhoon data released by ECMWF and NCEP were used as the true wind speed for the evaluation of wind measurement accuracy. Only wind speed data above 20 m/s during Typhoon Bavi were inversed here, because the training set in the Machine Learning method only includes data samples with wind speed greater than 20 m/s, as shown in Figure 4. A total of 7389 samples were available for the experiment over the four days. This subsection provides a detailed analysis of the CYGNSS satellite flight tracks and the corresponding true wind speeds during Typhoon Bavi.

Figure 5a shows the location of region for performance evaluation, and Figure 5b shows Typhoon Bavi (2020.8.22~2020.8.26) moving track map and daily area of interest. The CYGNSS data during 2020.8.22~2020.8.26 was first preprocessed for data, and then analyzed specifically according to time after obtaining analyzable data. Typhoon Bavi occurred in the western Pacific Ocean. The typhoon hourly track data used in this study was collected by Department of Water Resources of Zhejiang Province (http://typhoon.zjwater.gov.cn/, accessed on 20 June 2021). In addition, it was combined with the data distribution of CYGNSS to determine the specific typhoon area. Since there was no data in the region after preprocessing on 22 August 2020, this paper mainly studied the data from 23 August 2020 to 26 August 2020. In Figure 5b, the five pointed star represents the starting position of the typhoon, and the dotted box represents each divided typhoon area. Table 1 shows the specific selection range of each regional division.

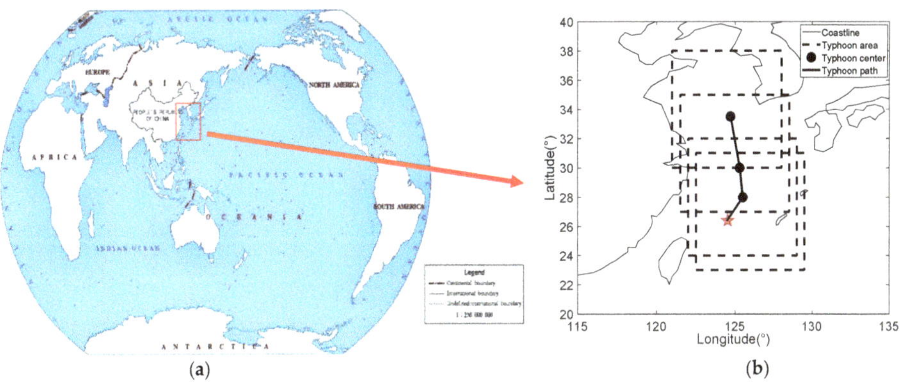

Figure 5. (**a**) Location of the region for performance evaluation (world map: preview number: GS (2016) 1563); (**b**) Typhoon Bavi (2020.8.22~2020.8.26) moving track map and daily interested area.

Table 1. 2020.8.22~2020.8.26 typhoon area latitude and longitude selection range.

Date	Longitude Range (°)	Latitude Range (°)
8.22	120°~127°	22°~30°
8.23	122.5°~129.5°	23°~31°
8.24	122°~129°	24°~32°
8.25	121.5°~128.5°	27°~35°
8.26	121°~128°	30°~38°

This paper mainly focused on the data with wind speed above 20 m/s. The proportion of samples has been determined in Section 2.3.2. Two measurement standards were used to compare the performance of three models: 1. Mean absolute error (MAE); 2. Root Mean Square Error (RMSE); and 3. Correlation Coefficient.

3.2. Analysis of Overall Inversion Results

The overall performance of the three trained models was investigated for all data during the typhoon, and Figure 6 shows the scatter plots of the true and inverse wind speed for all data during the typhoon for the three models. Table 2 shows the specific performance analysis of the three models.

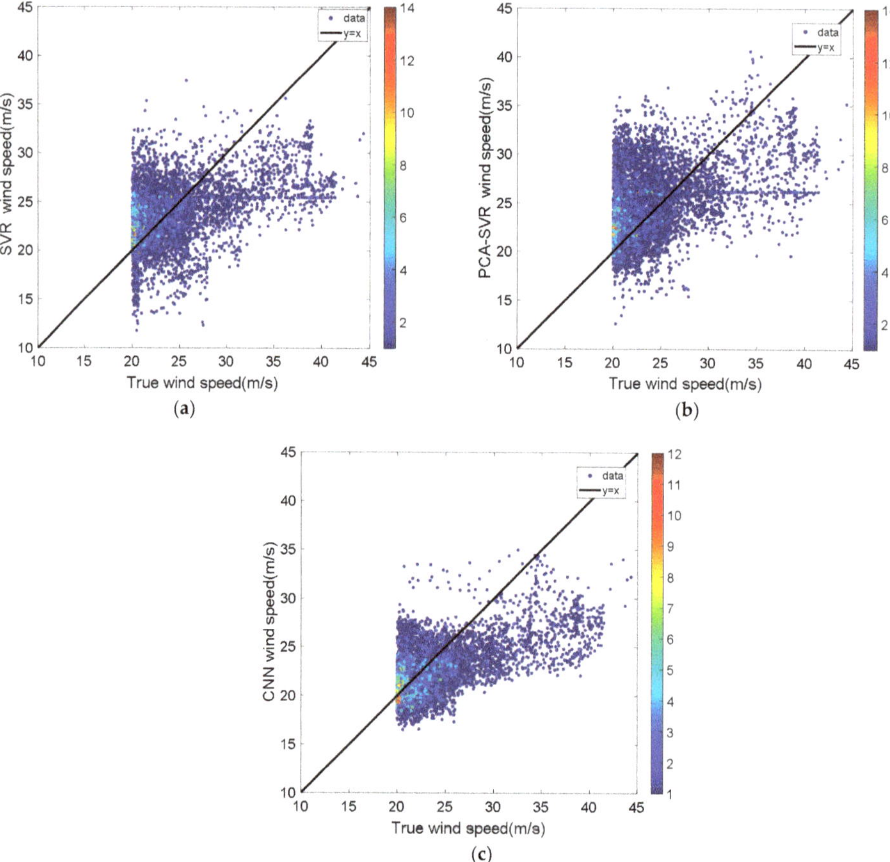

Figure 6. (**a**) SVR model wind speed inversion results; (**b**) PCA-SVR model wind speed inversion results; and (**c**) CNN model wind speed inversion results. The color bar on the right represents data density.

Table 2. Model performance analysis (Correl. Coef. represents the correlation coefficient).

Performance (m/s)	Overall Interval			20~30 m/s			Above 30 m/s		
	SVR	PCA-SVR	CNN	SVR	PCA-SVR	CNN	SVR	PCA-SVR	CNN
MAE	4.10	3.85	2.71	3.66	3.32	2.10	8.44	9.08	8.52
RMSE	5.48	5.10	3.80	4.88	4.17	2.64	9.51	10.50	9.22
Correl. Coef.	0.40	0.41	0.55	0.20	0.24	0.25	0.28	0.19	0.32

Firstly, it can be demonstrated from the scatter plot in Figure 6 that all three methods could be used to inverse the wind speed. In Figure 6, it was obvious that the SVR model inversion results had the greatest dispersion, and the inversion results reached a minimum of 10 m/s. The PCA-SVR model after adding data downscaling was partially improved for the problem of data divergence, but there was still a bias. The true wind speed of 20 m/s inversed the results around 35 m/s. While the CNN model had the most concentrated scattered data, the minimum inverse wind speed was about 15 m/s. The inversion results for the wind speed dataset around 20 m/s converged significantly and the outcomes were better than the other two methods. In general, the CNN method showed good inversion performance.

The performance of each of the three models was analyzed in three data intervals: (i) overall; (ii) 20~30 m/s; and (iii) above 30 m/s. From Table 2, except for the MAE value of CNN above 30 m/s, which was slightly inferior to SVR, all the error results indicated that CNN had the best performance. PCA-SVR was the second and SVR was the worst. Especially in the three data intervals, the correlation coefficients of CNN model were the highest. Further analysis showed that the MAE of CNN in the overall interval was improved by 33.90%, RMSE by 30.66% and correlation coefficient by 37.50% over SVR.

However, when the typhoon wind speed was greater than 30 m/s, the deviations of the wind speed values obtained from all three model inversions were all large, possibly because of the lack of higher wind speed train data (>40 m/s), as in Figure 4b, which leads to large bias in the inversion of typhoon data higher than 30 m/s.

3.3. Analysis of Daily Inversion Results by CNN Models

It was known from the analysis in Section 3.2 that the CNN model produced better wind speed inversion results for the overall data during typhoons. Considering the large variation of daily climatic environment and other factors during typhoons, which may affect the results of daily data collection from satellites for the same sea area, the CNN model was used for specific analysis of daily data. Figure 7 shows the daily CYGNSS satellite flight track and corresponding CNN wind speed, while Figure 8 corresponds to the absolute value of wind speed inversion (daily true wind speed minus the CNN model inverse wind speed). Table 3 shows the daily data performance results of the CNN model.

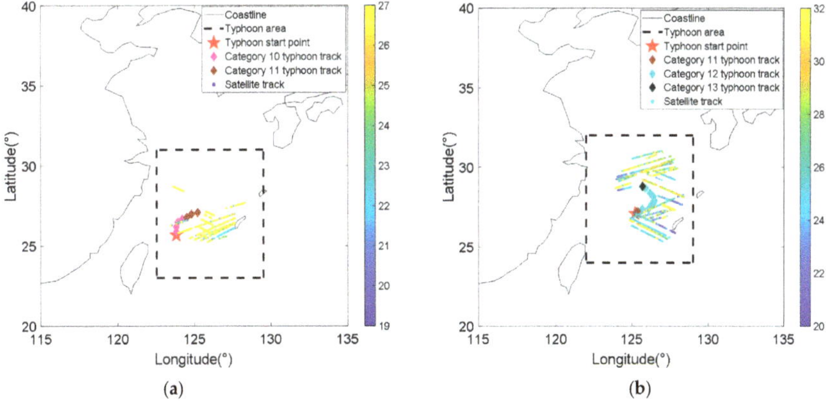

Figure 7. *Cont.*

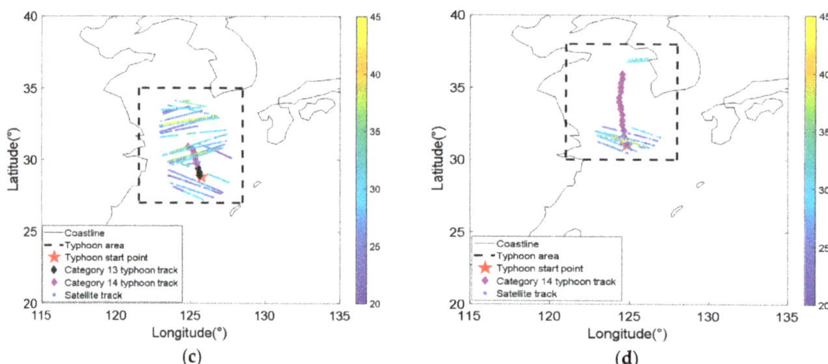

Figure 7. (**a**) 2020.8.23 CYGNSS satellite flight track and corresponding CNN wind speed; (**b**) 2020.8.24 CYGNSS satellite flight track and corresponding CNN wind speed; (**c**) 2020.8.25 CYGNSS satellite flight track and corresponding CNN wind speed; (**d**) 2020.8.25 CYGNSS satellite flight track and corresponding CNN wind speed. The color bar on the right represents the wind speed value, Unit: m/s.

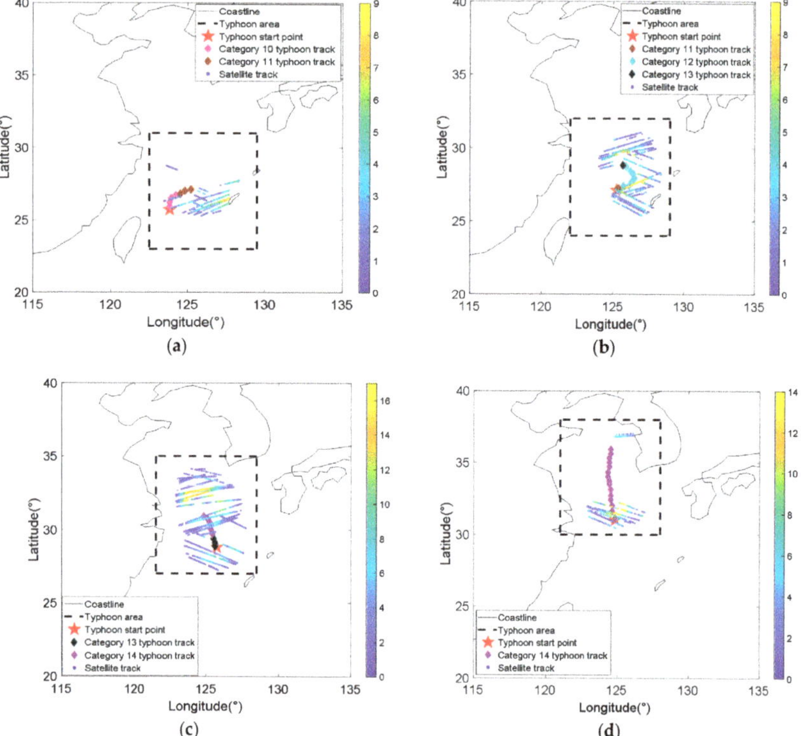

Figure 8. (**a**) 2020.8.23 CYGNSS satellite flight track and corresponding inversion error; (**b**) 2020.8.24 CYGNSS satellite flight track and corresponding inversion error; (**c**) 2020.8.25 CYGNSS satellite flight track and corresponding inversion error; (**d**) 2020.8.26 CYGNSS satellite flight track and corresponding inversion error. The color bar on the right represents the wind speed value, Unit: m/s.

Table 3. CNN model daily data performance analysis results.

Date	Aug. 23	Aug. 24	Aug. 25	Aug. 26
MAE (m/s)	2.33	2.29	4.18	4.21
RMSE (m/s)	2.95	2.92	5.70	5.25

It can be seen from Figure 7 that the CNN model could be used to inverse the typhoon wind speed, and the inverse wind speed can reach up to 55 m/s. Figure 8 and Table 3 show that the inversion results for 23 and 24 August 2020 were smaller errors compared to the last two days. The reason was that the true wind speeds of the first two days were mostly less than 30 m/s. The true wind speed of the data on 25 and 26 August 2020 was up to 45 m/s, and there were more data in the interval of 30 m/s to 45 m/s, so the CNN inversion results showed relatively large errors. This conclusion coincides with the results in Table 2.

The above contents have verified the accuracy of the model. Next, the comparison between the inverse wind speed and the typhoon track data was discussed. Table 4 shows the results of the comparison between CNN inverse wind speed, true wind speed (ECMWF and NCEP reanalysis wind speed data), and Beaufort scale of typhoon track data (from Department of Water Resources of Zhejiang Province). The approximate wind speed is similar to Best Track data. The CYGNSS samples here should meet less than spatial $\pm$ 0.5° and temporal $\pm$ 0.5 h from the typhoon track data. The five datasets satisfied the above conditions.

Table 4. Comparison results of wind speed data from Department of Water Resources of Zhejiang Province.

Beaufort Scale (Approximate Wind Speed)	**11 (30 m/s)**	**12 (33 m/s)**	**12 (38 m/s)**	**14 (42 m/s)**	**14 (42 m/s)**
Date	Aug. 24	Aug. 24	Aug. 25	Aug. 25	Aug. 26
True wind speed (m/s)	20.07	20.01	24.99	33.66	34.00
CNN wind speed (m/s)	19.24	22.88	27.47	28.94	24.78
Distance from the center of the typhoon track (km)	56.54	26.03	57.60	50.91	66.91

In Table 4, comparing with CNN wind speed and Typhoon track data, the first column result had the smallest deviation, and the fifth column result was the worst. It shows the greater wind speed level, the worse error is obtained. It is the same result as Tables 2 and 3, the reason has been analyzed before. However, in this paper, the true wind speed (ECMWF and NCEP reanalysis wind speed data) was used as the training benchmark of CNN model. As can be seen from Table 4, compared with the approximate wind speed (from Department of Water Resources of Zhejiang Province) during the typhoon, the true wind speed was actually underestimated, and the inversion performance of CNN model was limited by the true wind speed.

4. Conclusions

In response to the limitations of environmental conditions during typhoons, the high cost of collecting typhoon wind speed data leads to difficulties in obtaining training samples for high wind speeds. DDM observables such as DDMA and LES can change with the change of wind speed. Some traditional sea surface high wind speed inversion methods use a single DDM-derived observable (DDMA or LES), the incidence angle of specular reflection, and the significant wave height as parameters to establish GMF models with wind speed for wind speed inversion, which cannot fully explore the hidden features in the data. This limits the accuracy of high wind speed inversion. In order to use multi-dimensional data features to fully exploit the data during typhoons and improve the accuracy of the inversion of typhoon wind speeds in the sea area and the performance of real-time monitoring, this paper proposed a CYGNSS inversion method for high wind speed on the sea surface based on machine learning. Firstly, CYGNSS satellite data and true wind speed data from ECMWF and NCEP were used to construct the original datasets, and then three machine learning methods, SVR, PCA-SVR, and CNN, were used to train the data greater than 20 m/s during the typhoon. To avoid bias of the models, the undersampling method was adopted to control the number of samples. Lastly, the trained models were used for the inversion of Typhoon Bavi from 23 to 26 August 2020. The following conclusions could be drawn from the experimental results.

(1) All three models can be used to inverse the sea surface high wind speed from CYGNSS data. SVR can effectively solve the regression problem of high-dimensional characteristics, so the 27-dimensional characteristic parameters can be finally regressed to the wind speed value. Due to the large samples and high mapping dimension of kernel function, the calculation is too large, so PCA is used to reduce the dimension of data, which can speed up the training speed and obtain better wind speed inversion results.
(2) The CNN method can map arbitrarily complex nonlinear relationships and extract hidden deep-level features in the data. Even better, it also has the characteristics of strong robustness and self-learning capability. From an overall perspective, better results were obtained by using the CNN model for sea surface high wind speed inversion. The MAE of CNN was 2.71 m/s and RMSE was 3.8 m/s. Compared with the SVR model, the MAE of CNN was improved by 33.90% and RMSE improved by 30.66%. However, the inversion results of the three models for wind speeds above 30 m/s had large deviations. The reason for this error may be related to the lack of high wind speed data.
(3) The daily data inversion results during the typhoon show that CNN can be applied to the high wind speed inversion when the daily climate environment and other factors change greatly during the typhoon. Compared with the wind speed data at the typhoon center point provided by the Department of Water Resources of Zhejiang Province, it can be found that the higher the wind level, the larger the error between the true wind speed and the CNN inversion wind speed value near the typhoon center point. This error was caused by using underestimated true wind speeds (ECMWF and NCEP reanalysis wind speed data) to train the CNN model.

The difficulty of high wind speed inversion is the lack of higher wind speed samples, especially more than 40 m/s data, which leads to insufficient model training. Except for this, the selection of true wind speed during typhoons for training is also the key to the performance of the inversion. In the future, with the increasing amount of higher wind speed data and the use of more accurate model winds such as HWRF, GPS Dropsondes, and SFMR during typhoons, the accuracy of the obtained model will be improved and the error of typhoon inversion will be reduced. Eventually, the real-time prediction capability of typhoons will be realized.

Author Contributions: Y.Z. and Y.H. conceived and designed the framework of the study. J.Y. completed the data collection and processing. Y.Z. and J.Y. completed the algorithm design and the data analysis and were the lead authors of the manuscript, with contributions from J.Y., W.M., Z.Y. and S.Y. All authors have read and agreed to the published version of the manuscript.

Funding: This work was supported by the National Natural Science Foundation of China (Grant No. 41871325) and the National Key R&D Program of China (Project No. 2019YFD0900805).

Institutional Review Board Statement: Not applicable.

Informed Consent Statement: Not applicable.

Data Availability Statement: The raw/processed data required to reproduce these findings cannot be shared at this time as the data also forms part of an ongoing study.

Acknowledgments: Thanks NASA for the CYGNSSS public data; the European Centre for Medium-Range Weather Forecasts (ECMWF) and the National Centers for Environmental Prediction (NCEP) for the reanalysis dataset; and Department of Water Resources of Zhejiang Province for the typhoon track data. We would also like to thank Professor Yang Dongkai of Beijing University of Aeronautics and Astronautics and Li Weiqiang of CSIC-IEEC for their suggestions on GNSS-R satellite data analysis. We would like to thank Zhou Bo and Qin Jin from Shanghai Institute of Aerospace Electronics for their suggestions on the receiver of reflected signals.

Conflicts of Interest: The authors declare no conflict of interest.

References

1. Lin, M.; Sun, Y.; Zhen, S. Study on the inversion method of ocean wind field measurement by satellite-borne microwave scatterometer. *Acta Oceanol. Sin.* **1997**, *5*, 35–46.
2. Wang, Z.; Jiang, J.; Liu, J. Key technologies and scientific aspects of remote sensing of sea surface wind fields by all-polarization microwave radiometer. *Strateg. Stud. CAE* **2008**, *10*, 76–86. [CrossRef]
3. Zhang, W.; Shi, H.; Jiang, Z.; Yang, P.; Chang, S.; Xiang, J. Evaluation of variational scheme for synthetic aperture radar wind field inversion. *Chin. J. Geophys.* **2021**, *64*, 2436–2446. [CrossRef]
4. Hasager, C.B.; Hahmann, A.N.; Ahsbahs, T.; Karagali, I.; Sile, T.; Badger, M.; Mann, J. Europe's offshore winds assessed with synthetic aperture radar, ASCAT and WRF. *Wind Energy Sci.* **2020**, *5*, 375–390. [CrossRef]
5. Kilic, L.; Prigent, C.; Boutin, J.; Meissner, T.; English, S.; Yueh, S. Comparisons of Ocean Radiative Transfer Models With SMAP and AMSR2 Observations. *J. Geophys. Res. Oceans* **2019**, *124*, 7683–7699. [CrossRef]
6. Clarizia, M.; Ruf, C.; Jales, P. Spaceborne GNSS-R Minimum Variance Wind Speed Estimator. *IEEE Trans. Geosci. Remote Sens.* **2014**, *52*, 6829–6843. [CrossRef]
7. Clarizia, M.; Ruf, C. Wind Speed Retrieval Algorithm for the Cyclone Global Navigation Satellite System (CYGNSS) Mission. *IEEE Trans. Geosci. Remote Sens.* **2016**, *58*, 4419–4432. [CrossRef]
8. Wang, F.; Yang, D.; Zhang, B.; Li, W. Waveform-based spaceborne GNSS-R wind speed observation: Demonstration and analysis using UK TechDemoSat-1 data. *ADV Space Res.* **2018**, *61*, 1573–1587. [CrossRef]
9. Ruf, C.; Balasubramaniam, R. Development of the CYGNSS Geophysical Model Function for Wind Speed. *IEEE J-STARS* **2018**, *12*, 66–77. [CrossRef]
10. Reynolds, J.; Clarizia, M.; Santi, E. Wind Speed Estimation from CYGNSS Using Artificial Neural Networks. *IEEE J-STARS* **2020**, *13*, 708–716. [CrossRef]
11. Yang, D.; Liu, Y.; Wang, F. Research on the inversion method of satellite-based GNSS-R sea surface wind speed. *J. Electron. Inform. Technol.* **2018**, *40*, 462–469. [CrossRef]
12. Ruf, C.; Asharaf, S.; Balasubramaniam, R.; Gleason, S.; Lang, T.; McKague, D.; Twigg, D.; Waliser, D. InOrbit Performance of the Constellation of CYGNSS Hurricane Satellites. *Bull. Am. Meteorol. Soc.* **2019**. [CrossRef]
13. Rodriguez-Alvarez, N.; Garrison, J. Generalized Linear Observables for Ocean Wind Retrieval From Calibrated GNSS-R Delay–Doppler Maps. *IEEE Trans. Geosci. Remote* **2016**, *54*, 1142–1155. [CrossRef]
14. Said, F.; Katzberg, S.; Soisuvarn, S. Retrieving Hurricane Maximum Winds Using Simulated CYGNSS Power-Versus-Delay Waveforms. *IEEE J-STARS* **2017**, *10*, 3799–3809. [CrossRef]
15. Al-Khaldi, M.; Johnson, J.; Kang, Y.; Steven, J. Track-Based Cyclone Maximum Wind Retrievals Using the Cyclone Global Navigation Satellite System (CYGNSS) Mission Full DDMs. *IEEE J-STARS* **2019**, *13*, 21–29. [CrossRef]
16. Al-Khaldi, M.; Katzberg, S.J.; Johnson, J. Matched Filter Cyclone Maximum Wind Retrievals Using CYGNSS: Progress Update and Error Analysis. *IEEE J-STARS* **2021**, *99*, 3591–3601. [CrossRef]
17. Wu, C.; Yan, S.; Yang, Y.; Bu, F.; Chen, Z. An inversion method for ocean surface wind speed based on time-lapse-Doppler images. *Bull. Sci. Technol.* **2019**, *35*, 22–30. [CrossRef]
18. Gao, H.; Bai, Z.; Fan, D. GNSS-R sea surface wind speed inversion based on BP neural network. *Chin. J. Aeronaut* **2019**, *40*, 198–206. [CrossRef]
19. Wang, S. Research on GNSS-R Sea Surface Wind Speed Inversion Algorithm Based on Neural Network Model. Master's Thesis, University of Chinese Academy of Sciences, Beijing, China, 2020. [CrossRef]
20. Cardellach, E.; Nan, Y.; Li, W. Variational Retrievals of High Winds Using Uncalibrated CyGNSS Observables. *Remote Sens.* **2020**, *12*, 3930. [CrossRef]
21. Saïd, F.; Jelenak, Z.; Park, J.; Chang, P. The NOAA Track-Wise Wind Retrieval Algorithm and Product Assessment for CyGNSS. *IEEE Trans. Geosci. Remote Sens.* **2021**, 1–24. [CrossRef]
22. Shao, L.; Zhou, X.; Zhang, C.; Liu, H. Analysis of satellite-based GNSS-R typhoon observations. *Remote Sens. Inf.* **2020**, *4*, 35–39.
23. Gong, W.; Shi, C.; Zhang, T.; Meng, X. Evaluation of mean sea level pressure and surface wind speed from two numerical models in China. *J. Glaciol. Geocryol.* **2015**, *37*, 1497–1507. [CrossRef]
24. Hou, M.; Wang, G.; Bu, Q. Analysis of wind speed characteristics based on four reanalysis data offshore China. *Tianjin Sci. Technol.* **2017**, *44*, 109–113. [CrossRef]
25. Pan, Y.; Xu, J.; Zhang, Y.; Yuan, S.; Zhu, W. Simulation of the 2015 Northwest Pacific tropical cyclone based on the East Asian regional reanalysis system. *J. Zhanjiang Ocean Univ.* **2020**, *40*, 53–63.
26. Saini, J.; Dutta, M.; Marques, G. Fuzzy Inference System Tree with Particle Swarm Optimization and Genetic Algorithm: A novel approach for PM10 forecasting. *Syst. Appl.* **2021**, *183*, 115376. [CrossRef]

27. Bergadano, F.; Raedt, L. Machine Learning: ECML-94 Volume 784 | | Estimating attributes: Analysis and extensions of RELIEF. *LNCS* **1997**, *11*, 171–182. [CrossRef]
28. Last, M. Kernel Methods for Pattern Analysis. *J. Am. Stat. Assoc.* **2006**, *101*, 1730. [CrossRef]
29. Alex, J.; Bernhard, S. A tutorial on support vector regression. *Stat. Comput.* **2004**, *14*, 199–222. [CrossRef]
30. Liu, B.; Qi, X. Research on prediction of industrial solid waste generation in China based on PCA-SVR model. *J. Henan Normal Univ.* **2020**, *48*, 69–74. [CrossRef]
31. Ma, Q.; Zhang, X.; Zhang, C.; Zhou, H.; Wu, Z. Cross-wave velocity prediction based on one-dimensional convolutional neural network. *Lith Res.* **2021**, 1–10. Available online: http://kns.cnki.net/kcms/detail/62.1195.TE.20210530.1549.002.html (accessed on 5 June 2021).

MDPI
St. Alban-Anlage 66
4052 Basel
Switzerland
Tel. +41 61 683 77 34
Fax +41 61 302 89 18
www.mdpi.com

Remote Sensing Editorial Office
E-mail: remotesensing@mdpi.com
www.mdpi.com/journal/remotesensing

www.ingramcontent.com/pod-product-compliance
Lightning Source LLC
LaVergne TN
LVHW070625170726
843515LV00004B/177